SEEDS OF CHANGE
ADVANCE IN SUSTAINABLE FARMING

Tanishka Nale

Title   : Seeds of Change Advance in Sustainable Farming

Author   :   Tanishka Nale

Edition   :   First (September, 2024)

ISBN   :   9789348037817

Copyright © 2024, All Rights Reserved by Author

Published by

Regd. Add.: 254, Khuriyakhatta No. 10, Bindukhatta,
Lalkuan, Nainital - 262402, Uttarakhand, India
Website : www.taneeshapublishers.in
E-mail : taneeshapublishers@gmail.com
Phone : +91 845481 2712, +91 976041 7980

Printed by :

Manipal Technologies Limited, Bengaluru - 560001, Karnataka

INDEX

INDEX

A Brief History of Agriculture: An Indian Perspective

The story of agriculture in the Indian subcontinent is a rich tapestry woven with centuries of innovation, tradition, and resilience. From the verdant plains of the Indus Valley to the lush fields of the Ganges, agriculture has been the lifeblood of Indian civilization, sustaining communities and shaping the cultural landscape for millennia.

The roots of Indian agriculture can be traced back to the ancient Indus Valley civilization, one of the world's earliest urban societies. In the fertile floodplains of the Indus River, ancient peoples cultivated a diverse array of crops, including wheat, barley, rice, and cotton. The advanced urban centers of Mohenjo-Daro and Harappa boasted sophisticated agricultural techniques, including irrigation systems and crop rotation, which supported a thriving population and facilitated trade with distant lands.

With the decline of the Indus Valley civilization, agricultural practices continued to evolve across the Indian subcontinent, influenced by migrations, invasions, and the spread of new ideas and technologies. The arrival of Aryans brought with it the cultivation of crops like rice and millet, as well as the introduction of animal husbandry practices such as cattle rearing. Over time, these agricultural innovations became deeply ingrained in the cultural fabric of India, shaping dietary preferences, religious rituals, and social customs.

The Mauryan and Gupta empires, two of ancient India's most illustrious dynasties, fostered further advancements in agriculture, promoting land reforms, irrigation projects, and the dissemination of agricultural knowledge. The Gupta period, in particular, is often regarded as a golden age of Indian agriculture, marked by bountiful harvests, flourishing trade networks, and the spread of scientific farming techniques.

Throughout the medieval period, Indian agriculture continued to thrive, despite the political fragmentation and foreign invasions that characterized the era. The emergence of powerful agrarian kingdoms like the Cholas, Pallavas, and Vijayanagara Empire saw the expansion of agricultural land, the introduction of new crops such as sugarcane and spices, and the development of irrigation infrastructure like tanks and canals.

The colonial period ushered in significant changes to Indian agriculture, as British colonial policies prioritized the cultivation of cash crops like cotton, indigo, and opium for export to the West. This exploitative agrarian system, coupled with heavy taxation and land revenue policies, led to widespread poverty and rural unrest, culminating in the Indian independence movement of the 20th century.

Since gaining independence in 1947, India has made remarkable strides in agricultural development, harnessing scientific research, technological innovation, and government policies to increase food production and alleviate poverty in rural areas. The Green Revolution of the 1960s and

70s, which introduced high-yielding varieties of wheat and rice, transformed India from a food-deficient nation to one of the world's leading agricultural producers.

However, India's agricultural sector continues to face numerous challenges, including land degradation, water scarcity, and the impacts of climate change. As the nation strives to achieve food security for its burgeoning population, sustainable agriculture practices, equitable land distribution, and investments in rural infrastructure will be essential to ensuring a prosperous future for Indian farmers and communities.

The history of agriculture in India is a testament to the resilience, ingenuity, and cultural significance of farming communities across the subcontinent. From the ancient civilizations of the Indus Valley to the modern agricultural innovations of the 21st century, agriculture remains central to India's identity and prosperity. As the nation looks towards the future, it must continue to honor its agricultural heritage while embracing sustainable practices that ensure the well-being of both its people and the land they depend upon.

Overview of Traditional Farming Methods in India

Traditional farming methods in India are deeply rooted in the country's rich agricultural heritage, shaped by centuries of cultural practices, geographic diversity, and ecological wisdom. Across the vast expanse of the Indian subcontinent, farmers have employed a diverse array of techniques to cultivate crops and sustain their communities, drawing upon the land's natural resources and indigenous knowledge passed down through generations.

One of the most iconic features of traditional Indian agriculture is the use of organic and sustainable farming practices. Before the advent of chemical fertilizers and pesticides, Indian farmers relied on natural inputs such as compost, crop residues, and animal manure to nourish the soil and promote plant growth. This holistic approach to farming, often referred to as "green farming" or "zero-budget natural farming," emphasizes the importance of maintaining soil fertility, conserving water, and preserving biodiversity. Crop diversity has long been a hallmark of Indian agriculture, with farmers cultivating a wide range of traditional and indigenous varieties suited to local climates and soil conditions. In regions like the

Western Ghats and the Eastern Himalayas, agroforestry systems integrate trees, shrubs, and crops to enhance soil fertility, provide shade, and support biodiversity. Traditional crop rotations and intercropping practices help minimize pests and diseases while maximizing yields and nutrient cycling.

Water management has been another key aspect of traditional Indian farming, particularly in regions prone to drought and erratic rainfall. Ancient water harvesting structures such as tanks, ponds, and stepwells were constructed to capture and store rainwater, providing a reliable source of irrigation during dry seasons. In arid regions like Rajasthan, traditional farmers dug "khadins" or earthen embankments to trap rainwater and recharge groundwater aquifers, enabling agriculture in otherwise inhospitable landscapes.

Terracing is another traditional farming technique widely practiced in hilly and mountainous regions of India. In the Western Ghats of Kerala and Karnataka, for example, farmers terrace steep slopes to create flat terraces for rice cultivation, preventing soil erosion and maximizing arable land. These terraced fields, known as "padis" or "sawahs," are a testament to the ingenuity and adaptability of Indian farmers in harnessing the land's resources. Animal husbandry has been integral to traditional Indian agriculture, with farmers raising livestock such as cattle, goats, and buffalo for milk, meat, and traction. In regions like Punjab and Haryana, bullock carts are still used for plowing fields and transporting agricultural produce, preserving a centuries-old tradition of symbiotic relationship between humans and animals.

In addition to these agricultural practices, traditional farming in India is deeply intertwined with cultural rituals, festivals, and social customs. The celebration of festivals like Pongal in Tamil Nadu, Baisakhi in Punjab, and Onam in Kerala are deeply rooted in agrarian traditions, marking key milestones in the agricultural calendar and honoring the contributions of farmers to society.

While traditional farming methods in India have sustained rural communities for centuries, they are not without challenges in the modern era. Rapid urbanization, land degradation, and climate change pose existential threats to traditional agricultural systems, necessitating innovative solutions and policy interventions to ensure their continued viability. Traditional farming methods in India embody the wisdom, resilience, and sustainability of agrarian communities across the subcontinent. As the nation grapples with the complexities of modern agriculture, there is much to be learned from the traditional knowledge and practices that have sustained Indian farmers for generations. By embracing the principles of stewardship, diversity, and harmony with nature, India can chart a path towards a more resilient and sustainable agricultural future.

Introduction to the Need for Innovation in Agriculture: An Indian Perspective

For millennia, traditional farming methods have been the backbone of India's agricultural sector, nourishing vast populations and sustaining rural communities across the subcontinent. However, as the world's second-most populous country and with a population exceeding 1.3 billion people, India faces unprecedented challenges in meeting the food demands of its growing populace. In this context, the imperative for innovation in agriculture has never been more pressing.

Traditional farming methods, while effective in their time, are increasingly strained by a combination of factors including population growth, resource depletion, environmental degradation, and climate change. These methods, often reliant on intensive water usage, chemical fertilizers, and pesticides, are proving unsustainable in the long run. India's diverse agro-climatic zones, ranging from the fertile plains of Punjab to the arid regions of Rajasthan, present unique challenges that require innovative solutions tailored to local contexts.

Climate change poses one of the most significant threats to Indian agriculture, with rising temperatures, erratic rainfall patterns, and extreme weather events disrupting traditional cropping patterns and reducing yields. In states like Maharashtra and Karnataka, recurring droughts have devastated crops and exacerbated rural poverty, highlighting the urgent need for climate-resilient agricultural practices.

Soil degradation is another pressing concern, with overuse of chemical fertilizers and pesticides leading to nutrient depletion, erosion, and loss of soil fertility. In states like Punjab and Haryana, the overemphasis on rice-wheat monoculture has resulted in the depletion of groundwater levels and the degradation of soil health, posing long-term risks to agricultural productivity and sustainability.

Loss of biodiversity is yet another challenge facing Indian agriculture, as the widespread adoption of high-yielding crop varieties has led to the erosion of traditional seed diversity and the loss of indigenous crop varieties adapted to local conditions. The decline of pollinators such as bees and butterflies further threatens crop pollination and food security.

In the face of these challenges, there is a growing recognition of the need for innovation in agriculture to ensure the long-term viability and sustainability of India's food systems. This innovation encompasses a wide range of approaches, including technological advancements, policy reforms, and social innovations that promote ecological resilience, resource efficiency, and inclusivity.

Technological innovations such as precision agriculture, drip irrigation, and soil health monitoring systems hold promise for optimizing resource use, reducing environmental impact, and enhancing productivity. Biotechnological interventions such as genetically modified crops and biofortified varieties can help address challenges related to pests, diseases, and nutritional deficiencies, while also improving resilience to climate change. Policy reforms are also crucial for fostering an enabling environment for agricultural innovation, including incentives for sustainable farming practices, investment in agricultural research and extension services, and support for smallholder farmers, women, and marginalized communities. Initiatives such as the National Mission on Sustainable Agriculture (NMSA) and the Pradhan Mantri Krishi Sinchayee Yojana (PMKSY) aim to promote climate-resilient agriculture, water conservation, and soil health management.

Social innovations that empower farmers with knowledge, skills, and access to markets are equally important for driving agricultural innovation at the grassroots level. Farmer producer organizations (FPOs), digital agriculture platforms, and community-led initiatives like organic farming cooperatives enable smallholder farmers to adopt sustainable practices, access credit and markets, and build resilience to shocks and stresses.

The need for innovation in agriculture in India is both urgent and imperative in the face of mounting challenges posed by population growth, climate change, soil degradation, and loss of biodiversity. By embracing technological, policy, and social innovations, India can transform its agricultural sector into a resilient, sustainable, and inclusive engine of growth that ensures food security, enhances livelihoods, and protects the environment for future generations.

The Promise of Innovation in Agriculture

As we stand on the threshold of the 21st century, a remarkable wave of technological advancements is poised to revolutionize agriculture, offering unprecedented opportunities to address the pressing challenges facing food production and sustainability. From precision farming and hydroponics to genetic engineering and blockchain technology, a myriad of innovations holds the promise of transforming the agricultural landscape and shaping the future of food.

One of the most transformative innovations in agriculture is precision farming, which leverages advanced technologies such as satellite imagery, drones, and sensors to gather real-time data on soil conditions, weather patterns, and crop health. By analyzing this data, farmers can make informed decisions about planting, irrigation, fertilization, and pest management, optimizing inputs and maximizing yields while minimizing environmental impact. Precision farming not only increases productivity and efficiency but also reduces resource usage, making agriculture more sustainable and resilient in the face of climate change and resource constraints.

Vertical farming and hydroponic systems represent another promising innovation in agriculture, particularly in densely populated urban areas where arable land is scarce. These innovative cultivation methods involve growing crops in vertically stacked layers or nutrient-rich water solutions, allowing for year-round production of fresh fruits, vegetables, and herbs in controlled indoor environments. By minimizing the need for soil and water and eliminating dependence on weather conditions, vertical farming and hydroponics offer a sustainable solution to urban food security and promote local food production and self-sufficiency.

Biotechnology and genetic engineering hold immense promise for addressing the challenges of pest and disease resistance, climate resilience, and nutritional enhancement in crops. Through genetic modification, scientists can introduce desirable traits such as drought tolerance, disease resistance, and increased nutritional content into crop plants, enhancing their productivity and adaptability to changing environmental conditions. Genetically modified crops have the potential to reduce the need for chemical pesticides and fertilizers, lower production costs, and improve the nutritional quality of food, thus contributing to global food security and public health.

Blockchain technology, with its decentralized and transparent ledger system, is revolutionizing the way agricultural supply chains operate, ensuring traceability, transparency, and integrity from farm to fork. By recording and verifying every transaction and movement of agricultural products, blockchain enables consumers to make informed choices about the origin, quality, and sustainability of the food they consume. This technology not only enhances food safety and security but also fosters trust and accountability within the agricultural sector, promoting fair trade practices and ethical sourcing.

Despite the immense promise of these innovations, their widespread adoption and implementation face several challenges, including technological barriers, regulatory constraints, and socio-economic factors. Access to technology, particularly in rural and marginalized communities, remains a significant barrier to adoption, as does the high cost of investment and lack of technical expertise. Regulatory frameworks governing the use of biotechnology and genetic engineering vary widely across countries, posing obstacles to research, development, and commercialization. Socio-economic factors such as land tenure, access to credit, and market dynamics also influence the uptake of innovative agricultural practices, highlighting the importance of holistic and inclusive approaches to innovation.

The promise of innovation in agriculture is vast and multifaceted, offering solutions to some of the most pressing challenges facing food production and sustainability in the 21st century. From precision farming and vertical farming to biotechnology and blockchain technology, these innovations have the potential to revolutionize the way we produce, distribute, and consume food, ensuring a more resilient, efficient, and equitable agricultural system for generations to come. However, realizing this promise requires concerted efforts from governments, researchers, farmers, and industry stakeholders to overcome barriers and create an enabling environment for innovation to flourish. By harnessing the power of technology, science, and collaboration, we can build a more sustainable and food-secure future for all.

Embarking on a Journey of Agricultural Exploration

Within the following pages lies a captivating voyage into the vibrant realm of modern agriculture. Our expedition will traverse the frontiers of innovation, sustainability, and transformative practices propelling the evolution of food production. From the laboratories of research institutions to the humble fields of smallholder farmers, we shall unveil narratives of resilience, adaptability, and visionary thinking that sculpt the contours of tomorrow's agricultural vista. Come; accompany us as we set forth on this odyssey of discovery, unearthing the boundless potential of modern agriculture to nurture and preserve our planet for the prosperity of generations to come.

In our exploration, we will witness the fusion of tradition and technology, as ancient wisdom intersects with cutting-edge science to forge new pathways in farming. From vertical gardens in bustling cities to regenerative practices restoring depleted landscapes, each encounter will unveil the diversity and resilience of agricultural ecosystems. Together, let us embark on this journey with open minds and hopeful hearts, embracing the challenges and opportunities that lie ahead. For in the tapestry of modern agriculture, we find not just sustenance, but also a profound connection to the land and to each other. Join us as we navigate this fertile landscape, guided by curiosity, innovation, and a shared vision of a thriving future for all.

Precision Agriculture

Understanding Precision Agriculture Techniques

Precision agriculture, also referred to as precision farming or smart farming, epitomizes a transformative approach to agricultural production. At its essence, precision agriculture leverages advanced technology to optimize inputs, enhance productivity, and promote sustainability within farming operations. This methodology recognizes the inherent variability present in agricultural fields, encompassing differences in soil composition, moisture levels, nutrient distribution, and crop health. By discerning and comprehending this variability, farmers can tailor their management strategies to address the unique requirements of distinct areas within their fields.

The foundation of precision agriculture lies in the integration of hardware, software, and data analytics to attain precision and efficiency in farming practices. Key components of precision agriculture include Global Positioning System (GPS) technology, Geographic Information System (GIS) software, remote sensing tools, and data management platforms. These technologies work synergistically to facilitate precise mapping, monitoring, and decision-making across the agricultural landscape.

GPS technology serves as a cornerstone of precision agriculture by providing accurate spatial positioning data to farmers and agricultural machinery. With GPS-enabled devices installed on tractors, drones, and other farm equipment, farmers can precisely locate themselves within their fields, enabling precise navigation, mapping, and data collection. GPS technology also facilitates the creation of spatially referenced data layers, which form the basis for detailed field maps used in precision agriculture.

Geographic Information System (GIS) software plays a pivotal role in the analysis and interpretation of spatial data in precision agriculture. By integrating GPS-derived spatial information with additional data layers on soil characteristics, weather patterns, topography, and crop performance, GIS software enables farmers to generate comprehensive field maps that depict variability and patterns within agricultural landscapes. These maps serve as valuable decision-support tools, guiding farmers in optimizing resource allocation, implementing targeted interventions, and enhancing overall farm management practices.

Remote sensing technologies, including satellite imagery, aerial drones, and ground-based sensors, contribute valuable insights to precision agriculture by providing real-time and high-resolution data on crop health, soil moisture levels, and environmental conditions. By capturing multispectral imagery and other remote sensing data, farmers can monitor crop growth, detect pest infestations, and assess the efficacy of management practices with unprecedented accuracy and efficiency. These insights empower farmers to make timely and informed decisions, leading to improved crop yields, resource conservation, and profitability.

Data management platforms form the backbone of precision agriculture systems, facilitating the collection, storage, analysis, and visualization of vast amounts of agricultural data. These platforms integrate data from multiple sources, including GPS devices, GIS software, remote sensing tools, weather stations, and farm management software, into centralized databases accessible to farmers and agricultural stakeholders. By harnessing the power of big data analytics and machine learning algorithms, data management platforms enable farmers to derive actionable insights, optimize agronomic practices, and continuously improve farm productivity and sustainability.

In summary, precision agriculture represents a paradigm shift in agricultural production, harnessing technology and data-driven approaches to optimize resource use, enhance productivity, and promote sustainability. By leveraging GPS technology, GIS software, remote sensing tools, and data management platforms, farmers can gain valuable insights into field variability, make informed decisions, and adapt their management practices to suit the specific needs of their crops and soils.

As the agricultural sector continues to evolve, precision agriculture stands poised to revolutionize farming practices, ushering in a new era of efficiency, sustainability, and resilience for farmers and food systems worldwide.

Utilizing Drones and Satellite Imagery in Precision Agriculture

Remote sensing technology, encompassing drones and satellite imagery, has emerged as a powerful asset in the precision agriculture toolkit, revolutionizing the way farmers monitor, manage, and optimize their agricultural operations. Drones, also known as unmanned aerial vehicles (UAVs), equipped with sophisticated cameras and sensors, offer high-resolution imaging capabilities that enable farmers to capture detailed insights into crop health, nutrient status, and pest infestations from aerial vantage points. Simultaneously, satellite imagery provides a broader perspective, covering vast expanses of agricultural land with each pass, and detecting subtle variations in crop health, moisture levels, and soil composition. The synergy between drones and satellite imagery, complemented by advanced analytics, empowers farmers to make informed decisions, optimize resource allocation, and enhance overall farm productivity and sustainability.

Drones have emerged as versatile tools in precision agriculture, offering unparalleled capabilities for aerial imaging, mapping, and monitoring of agricultural landscapes. Equipped with high-resolution cameras, multispectral sensors, and thermal imaging technology, drones can capture detailed images and data on crop health, nutrient levels, water stress, and pest infestations with exceptional accuracy and efficiency. By conducting regular drone flights over agricultural fields, farmers can gain real-time insights into crop conditions, identify areas of concern or potential yield limitations, and take targeted action to address agronomic issues promptly.

Furthermore, drones offer significant advantages in terms of flexibility, accessibility, and cost-effectiveness compared to traditional ground-based methods or manned aircraft. With the ability to fly at varying altitudes and capture images from multiple angles, drones can provide comprehensive coverage of agricultural fields, even in remote or inaccessible areas. Moreover, drones can be deployed quickly and easily, allowing farmers to conduct frequent monitoring and assessment of field conditions throughout the growing season. The relatively low operating costs associated with drones make them an attractive investment for farmers seeking to improve their crop management practices and maximize yields.

Satellite imagery complements the capabilities of drones by providing a broader spatial perspective and covering larger geographic areas with each pass. Satellites equipped with multispectral sensors can capture data across multiple spectral bands, including visible, near-infrared, and thermal wavelengths, enabling the detection of subtle variations in crop health, moisture levels, and soil properties. By analyzing satellite imagery, farmers can gain valuable insights into field conditions, track changes in crop growth and development over time, and identify trends or patterns that may impact yield potential.

Additionally, satellite imagery offers the advantage of temporal consistency, allowing farmers to monitor crop performance and environmental conditions across entire growing seasons or even multiple years. By leveraging historical satellite data and time-series analysis techniques, farmers can identify long-term trends, assess the impact of management practices or environmental factors on crop productivity, and make informed decisions to optimize agronomic performance and mitigate risk. The integration of drone and satellite imagery with advanced analytics and decision support systems represents a transformative approach to precision agriculture, enabling farmers to harness the power of big data and artificial intelligence to drive operational efficiencies and achieve sustainable production outcomes. By leveraging the rich information provided by remote sensing technology, farmers can implement targeted interventions, optimize resource allocation, and mitigate risks, ultimately enhancing the resilience and productivity of agricultural systems in an increasingly complex and dynamic environment.

Benefits of Precision Agriculture: Resource Efficiency**

One of the primary benefits of precision agriculture is resource efficiency. By precisely targeting inputs such as water, fertilizers, and pesticides, farmers can reduce waste and minimize environmental impact. For example, variable rate application technology allows farmers to apply inputs at variable rates across their fields, based on site-specific conditions. This means that areas with high nutrient levels may receive less fertilizer, while areas with nutrient deficiencies may receive more, leading to more efficient use of resources and cost savings for farmers.

Precision agriculture also enables more precise irrigation management, helping to conserve water in regions facing water scarcity. Soil moisture sensors can monitor soil moisture levels in real-time, allowing farmers to irrigate only when necessary and avoid overwatering. By optimizing water usage, precision agriculture not only conserves a precious resource but also reduces energy consumption associated with pumping and distributing water.

Benefits of Precision Agriculture: Yield Optimization**

In addition to resource efficiency, precision agriculture offers significant potential for yield optimization. By identifying and addressing limiting factors such as soil acidity, nutrient deficiencies, and pest pressure, farmers can unlock the full genetic potential of their crops. For example, precision soil sampling and analysis enable farmers to tailor fertilizer applications to meet the specific needs of their crops, resulting in healthier plants and higher yields.

Furthermore, by monitoring crop health throughout the growing season, precision agriculture allows farmers to intervene early in the event of pest infestations or disease outbreaks. Timely interventions, such as targeted pesticide applications or crop rotations, can prevent yield losses and protect crop quality. The ability to respond quickly to emerging threats is particularly crucial in today's globalized food supply chain, where even small disruptions can have far-reaching consequences.

Harnessing the Power of Data

Central to the success of precision agriculture is the effective collection, analysis, and utilization of data. Modern farming operations generate vast amounts of data, from soil samples and weather records to drone imagery and equipment telemetry. By leveraging advanced analytics and machine learning algorithms, farmers can extract actionable insights from this data, guiding decision-making and driving continuous improvement.

Data-driven decision-making allows farmers to optimize their operations in real-time, responding dynamically to changing conditions and maximizing efficiency. For example, predictive analytics can forecast crop yields based on historical data and environmental factors, helping farmers to plan harvest schedules and manage storage and logistics accordingly. By harnessing the power of data, precision agriculture empowers farmers to make informed decisions that enhance productivity, profitability, and sustainability.

Overcoming Challenges and Barriers

While precision agriculture holds tremendous promise, its widespread adoption faces several challenges and barriers. One of the primary obstacles is the cost of implementation, including the initial investment in technology and the ongoing expenses associated with data management and analysis. Smallholder farmers, in particular, may lack access to the capital and resources needed to adopt precision agriculture practices, limiting their ability to reap the benefits of these technologies.

Furthermore, there are concerns surrounding data privacy and security, as well as issues related to interoperability and data standards. In an increasingly connected and data-driven agricultural landscape, it is essential to establish protocols and regulations to protect sensitive information and ensure compatibility between different systems and platforms. Collaboration and partnerships between stakeholders, including farmers, researchers, technology providers, and policymakers, will be critical to overcoming these challenges and driving the widespread adoption of precision agriculture.

Looking Ahead: The Future of Precision Agriculture

Despite the challenges, the future of precision agriculture appears bright, with continued advancements in technology and innovation driving the evolution of farming practices. As sensors become smaller, cheaper, and more sophisticated, and as connectivity improves, the potential for precision agriculture to transform global food production is greater than ever before. By harnessing the power of data, automation, and artificial intelligence, farmers can optimize their operations, minimize environmental impact, and ensure a sustainable food supply for generations to come.

Vertical Farming

Vertical farming represents a departure from traditional horizontal farming methods, aiming to maximize space utilization and resource efficiency. By stacking crops vertically in layers, vertical farms can achieve higher yields per square meter compared to conventional farming practices. Additionally, the controlled indoor environments of vertical farms allow for precise control over factors such as temperature, humidity, and light intensity, creating optimal conditions for plant growth year-round. This level of control not only increases productivity but also reduces the reliance on chemical inputs such as pesticides and herbicides, promoting environmentally sustainable agriculture. Moreover, vertical farming can be implemented in diverse urban settings, including rooftops, warehouses, and shipping containers, making it a versatile solution for urban food production. As India grapples with the dual challenges of urbanization and food security, vertical farming holds the potential to revolutionize the agricultural landscape and ensure a steady supply of fresh, locally grown produce for its growing population.

Principles of Vertical Farming

Vertical farming embodies a set of principles aimed at revolutionizing traditional agricultural practices by maximizing space utilization and optimizing growing conditions. In India, where land availability is at a premium, vertical farming presents a unique opportunity to achieve substantial yields within confined urban spaces. Through the implementation of innovative techniques such as hydroponics and aeroponics, Indian vertical farms can cultivate crops without soil, thus conserving valuable land resources while enhancing productivity. Furthermore, the integration of controlled environment agriculture (CEA) technologies empowers farmers to establish optimal growth conditions for plants, irrespective of external factors such as weather fluctuations or soil quality limitations.

At its core, the principle of maximizing space utilization drives the design and operation of vertical farms. Unlike traditional horizontal farming practices that spread crops across vast expanses of land, vertical farming capitalizes on vertical space by stacking crops in multiple layers. This vertical arrangement allows farmers to cultivate a greater number of plants within a smaller footprint, thereby increasing overall crop yields per unit area. In India, where rapid urbanization and population density pose significant constraints on available land for agriculture, the efficient use of vertical space holds immense potential for meeting the growing demand for fresh produce in urban areas.

Hydroponics and aeroponics represent two innovative techniques central to the practice of vertical farming in India. Hydroponics involves growing plants in nutrient-rich water solutions without the use of soil, while aeroponics entails suspending plant roots in a misty environment where nutrients are delivered through the air. Both methods offer distinct advantages in terms of resource efficiency, water conservation, and crop productivity. By eliminating the need for soil, hydroponic and aeroponic systems enable farmers to cultivate crops in controlled environments, free from soil-borne diseases and pests. Additionally, these systems allow for precise control over nutrient levels, pH balance, and moisture content, ensuring optimal plant growth and development.

Controlled environment agriculture (CEA) technologies play a pivotal role in supporting the principles of vertical farming by enabling farmers to create tailored growing conditions for crops. In India, where climatic variability and soil degradation pose significant challenges to traditional farming practices, CEA technologies offer a viable solution for ensuring consistent crop production year-round. By regulating factors such as temperature, humidity, light intensity, and carbon dioxide levels, CEA systems create ideal microclimates that promote optimal plant growth and minimize environmental stressors. This level of control not only enhances crop quality and yield but also reduces the reliance on chemical inputs such as pesticides and fertilizers, thereby promoting sustainable agriculture practices.

The principles of vertical farming underscore a paradigm shift in agricultural production towards sustainability, efficiency, and resilience. By maximizing space utilization, adopting innovative cultivation techniques such as hydroponics and aeroponics, and leveraging controlled environment agriculture technologies, Indian vertical farms can overcome the constraints of limited land availability and environmental variability to achieve high yields in small urban spaces. As India continues to urbanize and its population grows, vertical farming holds the potential to revolutionize the agricultural landscape,

Advantages of Vertical Farming: Year-Round Production**

In India, the agricultural sector faces numerous challenges, including seasonal variations and the adverse effects of climate change, which often disrupt traditional farming practices and affect crop yields. However, vertical farming offers a promising solution by enabling year-round production within controlled indoor environments. By leveraging advanced technologies and innovative cultivation methods, Indian vertical farms can maintain consistent growing conditions, ensuring a steady supply of fresh produce irrespective of external climatic fluctuations.

The controlled indoor environments of vertical farms provide a shield against the vagaries of weather, enabling farmers to regulate factors such as temperature, humidity, and light intensity to suit the needs of different crops. This level of control minimizes the risks associated with extreme weather events, pests, and diseases, thereby safeguarding crop yields and ensuring reliable production throughout the year. Additionally, vertical farming allows for the cultivation of a wide variety of crops, including leafy greens, herbs, fruits, and vegetables, regardless of their natural growing seasons, further enhancing the diversity and availability of fresh produce in the market.

The reliability of year-round production offered by vertical farming is particularly crucial for meeting the nutritional needs of India's rapidly growing population. With urbanization on the rise and dietary preferences shifting towards fresh and healthy foods, the demand for locally grown produce has surged. Vertical farming addresses this demand by providing a consistent supply of high-quality, nutritious produce that meets the preferences of consumers year-round. By reducing dependence on imported or out-of-season produce, vertical farming strengthens food security and sovereignty, ensuring access to fresh and affordable food for all segments of society.

Furthermore, year-round production through vertical farming supports the growth of India's agricultural economy by creating new opportunities for farmers and entrepreneurs. The controlled environment of vertical farms allows for precise management of resources, optimizing resource use efficiency and reducing production costs. This efficiency translates into higher profitability for farmers, enabling them to generate steady income streams throughout the year. Additionally, the scalability of vertical farming enables farmers to expand their operations and diversify their crop portfolios, thereby increasing their resilience to market fluctuations and economic uncertainties.

The advantages of year-round production offered by vertical farming are manifold, encompassing food security, economic prosperity, and environmental sustainability. By maintaining consistent growing conditions within controlled indoor environments, Indian vertical farms can ensure a steady supply of fresh produce, regardless of external climatic variations. This reliability strengthens food security, supports economic growth, and promotes sustainable agricultural practices, positioning vertical farming as a transformative force in India's agricultural landscape ensuring a steady supply of fresh, locally grown produce for its burgeoning urban population.

Advantages of Vertical Farming: Resource Efficiency

Resource efficiency stands as a cornerstone of vertical farming, offering significant advantages in a country like India, where water scarcity and energy constraints pose formidable challenges to traditional agriculture. Vertical farming techniques, notably hydroponics and aeroponics, epitomize resource-efficient cultivation methods that minimize water usage while maximizing crop productivity. These techniques represent sustainable alternatives to conventional soil-based agriculture, particularly in India's water-stressed regions.

Hydroponics, a key component of vertical farming, involves cultivating plants in nutrient-rich water solutions without the need for soil. By directly delivering essential nutrients to plant roots, hydroponic systems eliminate the inefficiencies associated with soil-based agriculture, where water is often wasted through runoff and evaporation. Instead, hydroponic farms recirculate water within closed-loop systems, drastically reducing water consumption compared to traditional farming methods. This water-saving feature is especially crucial in India, where agriculture accounts for a significant portion of freshwater usage, and water scarcity poses a growing threat to agricultural sustainability.

Similarly, aeroponics, another vertical farming technique, suspends plant roots in a misty environment where nutrients are delivered through the air. By exposing plant roots to a fine mist of nutrient solution, aeroponic systems maximize nutrient uptake efficiency while minimizing water usage. This innovative approach to crop cultivation eliminates the need for soil altogether, further reducing water requirements and enhancing resource efficiency. Moreover, aeroponic systems promote rapid plant growth and development, resulting in higher yields with minimal resource inputs.

In addition to water efficiency, vertical farming offers opportunities for energy savings and renewable energy integration, further enhancing resource efficiency in agricultural production. Indian vertical farms can harness renewable energy sources such as solar power to meet their energy needs, thereby reducing their environmental footprint and reliance on fossil fuels. Solar panels installed on rooftops or integrated into vertical farm structures can generate clean, renewable electricity to power lighting, climate control systems, and other energy-intensive operations. This shift towards renewable energy adoption not only reduces greenhouse gas emissions but also enhances the sustainability and resilience of vertical farming operations in India.

The resource efficiency advantages of vertical farming, particularly in terms of water conservation and renewable energy utilization, make it a compelling solution for addressing India's agricultural sustainability challenges. By implementing hydroponic and aeroponic techniques and integrating renewable energy sources, Indian vertical farms can achieve higher yields with fewer resources, contributing to food security, environmental conservation, and economic prosperity. As India strives to overcome water scarcity and energy constraints, vertical farming stands poised to play a pivotal role in shaping a more sustainable and resilient agricultural future.

Case Study: UrbanKisaan

UrbanKisaan, based in Hyderabad, India, is a pioneering vertical farming startup that aims to revolutionize urban agriculture in the country. Founded in 2017, UrbanKisaan utilizes hydroponic and aquaponic systems to grow a variety of leafy greens, herbs, and microgreens in vertically stacked growing towers. By leveraging technology and innovation, UrbanKisaan maximizes space utilization and minimizes resource usage, making urban farming accessible and sustainable in India's rapidly growing cities.

One of UrbanKisaan's flagship projects is located in the heart of Hyderabad, inside a converted warehouse spanning over 10,000 square feet. This vertical farm produces over 20 varieties of fresh produce year-round, supplying local restaurants, grocery stores, and consumers with locally grown, pesticide-free vegetables and herbs. By promoting urban farming as a viable alternative to traditional agriculture, UrbanKisaan is paving the way for a more sustainable and resilient food system in India.

Case Study: Future Farms

Future Farms, headquartered in Bengaluru, India, is a leading provider of vertical farming solutions for commercial agriculture. Founded in 2017, Future Farms specializes in designing and building turnkey vertical farming systems that enable farmers to grow a wide range of crops in controlled indoor environments. By integrating cutting-edge technologies such as IoT sensors, automated climate control, and data analytics, Future Farms empowers farmers to optimize their operations and maximize yields.

One of Future Farms' notable projects is located in Pune, Maharashtra, where they have partnered with local farmers to establish a vertical farming facility spanning over 50,000 square feet. This state-of-the-art farm utilizes vertical growing towers and advanced lighting systems to produce a variety of high-value crops, including strawberries, cherry tomatoes, and bell peppers. By providing farmers with the tools and knowledge to succeed in vertical farming, Future Farms is catalyzing the adoption of sustainable agriculture practices across India.

Case Study: Sky Greens

Sky Greens, based in Singapore, is an innovative vertical farming company that has expanded its operations to India with the aim of addressing food security challenges in urban areas. Founded in 2010, Sky Greens specializes in the development of patented vertical farming technology known as the A-Go-Gro system. This innovative system utilizes rotating tiers to maximize sunlight exposure and airflow, resulting in higher yields and lower energy consumption compared to traditional vertical farming methods.

In 2019, Sky Greens established its first vertical farm in India, located in Bengaluru, Karnataka. This state-of-the-art facility utilizes Sky Greens' A-Go-Gro system to produce a variety of leafy greens and vegetables for local markets and retailers. By leveraging technology and expertise from Singapore, Sky Greens is helping to build a more sustainable and resilient food system in India, one vertical farm at a time.

Introduction to Hydroponics and Aquaponics

Hydroponics and aquaponics stand at the forefront of modern agricultural innovation, offering sustainable alternatives to traditional soil-based farming practices. These systems have garnered widespread attention for their ability to cultivate crops efficiently, conserve resources, and maximize yields, making them integral components of the evolving agricultural landscape. In this chapter, we embark on a journey to uncover the principles underpinning hydroponic and aquaponic cultivation, examine their distinctions from conventional farming methods, and explore the diverse array of crops that flourish within these innovative environments.

Hydroponics: Cultivating Without Soil

Hydroponics represents a revolutionary approach to crop cultivation, eschewing the need for soil entirely and instead delivering essential nutrients directly to plant roots through nutrient-rich water solutions. This soil-less cultivation method allows for precise control over nutrient levels, pH balance, and water delivery, optimizing plant growth and development while minimizing resource wastage. By harnessing hydroponic systems, farmers can cultivate crops in diverse environments, ranging from urban rooftops to arid desert landscapes, transcending the limitations imposed by soil quality and climatic variability.

Hydroponics represents a revolutionary approach to crop cultivation, eschewing the need for soil entirely and instead delivering essential nutrients directly to plant roots through nutrient-rich water solutions. This soil-less cultivation method allows for precise control over nutrient levels, pH balance, and water delivery, optimizing plant growth and development while minimizing resource wastage. By harnessing hydroponic systems, farmers can cultivate crops in diverse environments, ranging from urban rooftops to arid desert landscapes, transcending the limitations imposed by soil quality and climatic variability.

Hydroponics represents a revolutionary approach to crop cultivation, eschewing the need for soil entirely and instead delivering essential nutrients directly to plant roots through nutrient-rich water solutions. This soil-less cultivation method allows for precise control over nutrient levels, pH balance, and water delivery, optimizing plant growth and development while minimizing resource wastage. By harnessing hydroponic systems, farmers can cultivate crops in diverse environments, ranging from urban rooftops to arid desert landscapes, transcending the limitations imposed by soil quality and climatic variability.

The principles of hydroponics revolve around maximizing resource efficiency and crop productivity through innovative cultivation techniques. Various hydroponic systems, including nutrient film technique (NFT), deep water culture (DWC), and aeroponics, offer distinct advantages in terms of water conservation, nutrient uptake efficiency, and space utilization. These systems enable farmers to achieve higher yields with fewer resources, paving the way for sustainable agriculture practices that mitigate environmental impact and promote food security.

Aquaponics: Symbiotic Farming Ecosystems

Aquaponics represents a symbiotic integration of aquaculture and hydroponics, harnessing the natural processes of fish farming and plant cultivation to create self-sustaining ecosystems. In aquaponic systems, fish and plants coexist in a mutually beneficial relationship, with fish waste serving as a nutrient source for plants, and plants acting as natural filters to purify the water for fish. This closed-loop cycle of nutrient exchange enables farmers to cultivate both aquatic species and terrestrial crops in a single integrated system, maximizing resource utilization and productivity. The core principle of aquaponics lies in harnessing the biological interactions between fish, plants, and beneficial microorganisms to create harmonious ecosystems that mimic natural aquatic environments. By capitalizing on the nutrient-rich waste produced by fish, aquaponic systems eliminate the need for synthetic fertilizers, reducing environmental pollution and enhancing sustainability. Moreover, the dual production of fish and crops within aquaponic systems offers opportunities for diversification and resilience, enabling farmers to optimize economic returns while minimizing input costs.

Comparing with Traditional Farming Methods

In contrast to traditional soil-based farming methods, hydroponics and aquaponics offer several distinct advantages in terms of resource efficiency, crop quality, and environmental sustainability. Traditional farming practices often rely on extensive land clearance, chemical inputs, and intensive water usage, leading to soil degradation, water pollution, and habitat destruction. In contrast, hydroponic and aquaponic systems require minimal land and water resources, utilize organic nutrient sources, and promote ecological balance, resulting in higher yields and lower environmental impact.

Exploring Crop Diversity in Hydroponic and Aquaponic Systems

One of the most intriguing aspects of hydroponic and aquaponic cultivation is the remarkable diversity of crops that thrive within these innovative environments. While traditional soil-based farming imposes limitations based on soil type, climate, and seasonal variations, hydroponic and aquaponic systems offer greater flexibility in crop selection and cultivation. A wide range of crops, including leafy greens, herbs, fruits, and vegetables, can be successfully grown in hydroponic and aquaponic systems, providing farmers with opportunities for year-round production and market diversification.

Hydroponics and aquaponics represent groundbreaking approaches to agricultural cultivation, offering efficient, sustainable, and resilient alternatives to traditional farming methods. By harnessing innovative technologies and ecological principles, these systems enable farmers to maximize resource utilization, optimize crop yields, and mitigate environmental impact, thereby shaping a more sustainable and food-secure future. Through further exploration and adoption of hydroponic and aquaponic systems, we can unlock the full potential of modern agriculture to nourish and sustain our growing global population.

Explanation of Hydroponic and Aquaponic Systems

Hydroponic systems represent a revolutionary approach to agriculture, fundamentally transforming the way plants are cultivated by eliminating the need for soil. Instead, plants are grown in nutrient-rich water solutions, allowing their roots to access essential nutrients directly. In hydroponic setups, inert mediums such as perlite, coconut coir, or rockwool provide support for the plants while allowing for efficient water and nutrient uptake. By carefully monitoring and adjusting nutrient solutions, farmers can ensure that plants receive the optimal balance of essential elements necessary for robust growth and development.

Aquaponic systems take the concept of hydroponics a step further by integrating it with aquaculture, creating symbiotic ecosystems where fish and plants coexist harmoniously. In aquaponic setups, fish are raised in tanks, with their waste serving as a valuable nutrient source for the plants. As fish excrete ammonia-rich waste, beneficial bacteria convert it into nitrites and nitrates, essential nutrients for plant growth. The nutrient-rich water is then circulated to the hydroponic growing beds, where plants absorb the nutrients, effectively filtering and purifying the water for the fish. This closed-loop cycle of nutrient exchange creates a self-sustaining ecosystem where fish and plants thrive in mutual benefit.

Hydroponic systems offer several distinct advantages over traditional soil-based agriculture. By eliminating the need for soil, hydroponic setups minimize the risk of soil-borne diseases and pests, resulting in healthier plants and higher yields. Additionally, the controlled environment of hydroponic systems allows for precise control over growing conditions, including nutrient levels, pH balance, and water quality, leading to optimized plant growth and resource utilization. Moreover, hydroponic cultivation requires significantly less water compared to traditional farming methods, making it particularly suitable for regions prone to water scarcity or drought. Aquaponic systems leverage the benefits of both hydroponics and aquaculture, creating synergistic relationships between fish and plants that enhance overall productivity and sustainability. In addition to providing a nutrient source for plants, fish in aquaponic systems also serve as a valuable protein source, further diversifying the potential outputs of the system. Moreover, the integration of fish and plants in aquaponic setups promotes ecological balance and resource efficiency, making it a compelling solution for sustainable food production in both rural and urban environments.

Hydroponic and aquaponic systems represent innovative approaches to agriculture that offer efficient, sustainable, and environmentally friendly alternatives to traditional farming methods. By harnessing the power of water and nutrients, these systems enable farmers to cultivate crops with higher yields, reduced resource inputs, and minimal environmental impact, thereby paving the way for a more resilient and food-secure future. Through further research, innovation, and adoption, hydroponic and aquaponic systems hold the potential to revolutionize global food production and contribute to a more sustainable and equitable food system for all.

Comparison with Traditional Soil-Based Farming

The adoption of hydroponic and aquaponic systems in agriculture represents a paradigm shift away from traditional soil-based farming practices, offering a host of advantages that address critical challenges facing modern agriculture. In comparison to conventional methods, hydroponics and aquaponics demonstrate superior resource efficiency, environmental sustainability, and productivity, making them compelling alternatives for sustainable food production.

One of the most significant advantages of hydroponic and aquaponic systems lies in their significantly reduced water usage compared to traditional soil-based farming. In conventional agriculture, water is often applied indiscriminately, leading to substantial wastage through evaporation, runoff, and deep percolation. In contrast, hydroponic and aquaponic setups utilize closed-loop systems where water is recirculated and reused, resulting in water savings of up to 90%. This water-efficient approach not only conserves precious freshwater resources but also mitigates the environmental impacts of agricultural water usage, such as soil erosion, nutrient leaching, and water pollution.

Moreover, the elimination of soil in hydroponic and aquaponic systems offers several distinct advantages over traditional farming methods. Soil quality varies widely across different regions, with many areas suffering from poor soil fertility, contamination, or erosion. By circumventing the need for soil altogether, hydroponic and aquaponic systems overcome these limitations, allowing crops to thrive in controlled environments optimized for growth. This soil-less cultivation approach is particularly advantageous in regions with limited arable land or urban settings where soil availability is scarce, enabling food production in unconventional locations such as rooftops, warehouses, or vertical farms.

Furthermore, hydroponic and aquaponic systems enable precise control over environmental factors such as temperature, humidity, and nutrient levels, leading to accelerated growth rates and higher yields compared to traditional farming methods. In soil-based agriculture, farmers have limited control over these variables, resulting in slower growth rates, lower yields, and increased susceptibility to weather-related stresses, pests, and diseases. In contrast, hydroponic and aquaponic setups provide optimal growing conditions tailored to the specific needs of plants, resulting in enhanced plant health, vigor, and productivity throughout the growing cycle. Additionally, the controlled environments of hydroponic and aquaponic systems facilitate year-round production, independent of seasonal variations or climatic constraints. Traditional farming methods are often subject to the vagaries of weather, leading to seasonal fluctuations in crop availability and quality. By operating within controlled indoor environments, hydroponic and aquaponic farms can maintain consistent growing conditions, ensuring a steady supply of fresh produce throughout the year. This reliability is essential for meeting the demands of consumers and markets for locally grown, high-quality produce, reducing reliance on imported or out-of-season crops.

The comparison between hydroponic/aquaponic systems and traditional soil-based farming underscores the transformative potential of these innovative cultivation methods. Through their superior resource efficiency, environmental sustainability, and productivity, hydroponic and aquaponic systems offer a pathway to a more resilient, equitable, and sustainable food system for the future. As global challenges such as population growth, climate change, and water scarcity continue to escalate, the adoption of hydroponic and aquaponic technologies holds promise for addressing these pressing issues and ensuring food security and sovereignty for generations to come.

Examples of Crops Grown Using Hydroponics and Aquaponics

Hydroponic and aquaponic systems offer versatile platforms for the cultivation of a diverse range of crops, spanning from leafy greens and herbs to fruiting vegetables and even certain fruit trees. These innovative cultivation methods provide optimal growing conditions, precise nutrient delivery, and efficient resource utilization, resulting in robust plant growth and high-quality yields. Let us explore some of the exemplary crops that flourish within hydroponic and aquaponic environments, showcasing the breadth and potential of these sustainable farming systems.

Leafy Greens:

Leafy greens such as lettuce, spinach, kale, and Swiss chard are among the most popular and well-suited crops for hydroponic cultivation. These greens have shallow root systems and high water requirements, making them ideally suited for hydroponic systems where nutrient-rich water solutions can be efficiently delivered directly to their roots. In hydroponic setups, leafy greens exhibit rapid growth rates, vibrant foliage, and exceptional nutritional profiles, making them prized commodities in both consumer and commercial markets. The controlled environment of hydroponic systems ensures optimal growing conditions, resulting in crisp textures, vibrant colors, and prolonged shelf life for leafy greens.

Herbs

Herbs are another category of crops that thrive in hydroponic environments, producing flavorful and aromatic leaves prized in culinary applications. Basil, cilantro, mint, and parsley are among the most commonly cultivated herbs in hydroponic systems, offering a continuous supply of fresh, high-quality foliage year-round. Hydroponic cultivation promotes vigorous growth and robust flavor development in herbs, as plants receive precise nutrient doses and consistent environmental conditions conducive to optimal growth. Moreover, hydroponic herbs boast extended shelf life and superior flavor intensity compared to their soil-grown counterparts, making them sought-after ingredients in restaurants, kitchens, and markets worldwide.

Fruiting Vegetables

While leafy greens and herbs dominate hydroponic cultivation, certain fruiting vegetables also thrive in these innovative systems, particularly those with compact growth habits and shallow root systems. Tomatoes, cucumbers, peppers, and strawberries are exemplary examples of fruiting vegetables amenable to hydroponic cultivation. In hydroponic setups, these crops exhibit accelerated growth rates, increased fruit yields, and enhanced fruit quality compared to traditional soil-based farming methods. The controlled nutrient delivery and environmental parameters of hydroponic systems contribute to the development of robust root systems, prolific flowering, and bountiful fruiting in these vegetables, resulting in flavorful, nutrient-dense harvests prized by consumers.

Aquaponic Systems

Aquaponic systems offer a unique integration of hydroponics with aquaculture, creating symbiotic ecosystems where fish and plants coexist harmoniously. In aquaponic setups, crops such as tomatoes, cucumbers, peppers, and strawberries can be cultivated alongside fish species like tilapia, catfish, and trout. These fruiting vegetables thrive in the nutrient-rich water provided by the aquaculture component, benefiting from the organic nutrients generated by fish waste and the biological filtration provided by plant roots. The integration of fish and plants in aquaponic systems creates self-sustaining ecosystems that maximize resource utilization, promote ecological balance, and yield abundant harvests of both crops and fish, showcasing the potential for diversified and sustainable food production.

The examples of crops grown using hydroponics and aquaponics demonstrate the versatility, productivity, and sustainability of these innovative cultivation methods. From leafy greens and herbs to fruiting vegetables and beyond, hydroponic and aquaponic systems offer viable solutions for meeting the diverse dietary needs of a growing global population while minimizing environmental impact and maximizing resource efficiency. As the adoption of hydroponic and aquaponic technologies continues to expand, so too will the range of crops cultivated within these innovative farming systems, heralding a new era of sustainable agriculture for the future.

Urban Farming

Urban Farming Initiatives: Cultivating Sustainable Solutions in Urban Landscapes

Introduction

Urban farming, also referred to as urban agriculture, is a multifaceted approach to addressing various challenges faced by urban areas, ranging from food insecurity to environmental sustainability. This paper provides an overview of urban farming initiatives, highlighting their significance, objectives, and impact on urban communities.

The Significance of Urban Farming Initiatives

Urban farming initiatives have garnered increasing attention due to their potential to address pressing urban issues. With the global population becoming increasingly urbanized, cities are faced with challenges such as limited access to fresh produce, food deserts, and environmental degradation. Urban farming offers a sustainable solution by utilizing underutilized spaces within cities to grow food, thereby reducing reliance on long-distance food transportation and mitigating the environmental footprint associated with conventional agriculture.

Objectives of Urban Farming Initiatives

Urban farming initiatives pursue several objectives, including:

1. **Food Security:** By producing food locally, urban farming initiatives aim to increase access to fresh, nutritious produce, particularly in underserved communities where access to supermarkets or grocery stores is limited.
2. **Environmental Sustainability:** Urban farming promotes sustainable agriculture practices, such as organic farming methods and water conservation techniques, to minimize environmental impact and preserve natural resources.
3. **Community Engagement:** Through community gardens, rooftop farms, and farmers' markets, urban farming initiatives foster community engagement and social cohesion, providing opportunities for education, skill-building, and social interaction.
4. **Economic Opportunities:** Urban farming creates employment opportunities and supports local economies by generating income for small-scale farmers, entrepreneurs, and food-related businesses.

Types of Urban Farming Initiatives

Urban farming initiatives encompass a diverse range of activities and approaches, including:

1. **Rooftop Gardens:** Utilizing rooftops of buildings for gardening purposes, rooftop gardens maximize space efficiency and contribute to urban greening efforts while providing fresh produce to residents.
2. **Community Gardens:** Community gardens are shared spaces where individuals or groups collectively cultivate fruits, vegetables, and herbs, promoting community bonding and healthy eating habits.
3. **Vertical Farms:** Vertical farming utilizes vertical space in buildings or structures to grow crops hydroponically or aeroponically, enabling year-round production in urban environments with limited land availability.
4. **Aquaponic Systems:** Aquaponics combines aquaculture (fish farming) with hydroponics (soilless plant cultivation), creating a symbiotic ecosystem where fish waste provides nutrients for plants, and plants filter water for fish, resulting in a closed-loop system that

minimizes water usage.

Impact of Urban Farming Initiatives

Urban farming initiatives have demonstrated significant impact on urban communities:

1. **Improved Access to Fresh Produce:** By bringing food production closer to consumers, urban farming initiatives increase access to fresh, locally grown produce, contributing to improved dietary choices and public health outcomes.

2. **Environmental Benefits:** Urban farming practices such as composting, rainwater harvesting, and green infrastructure contribute to environmental sustainability by reducing greenhouse gas emissions, mitigating urban heat island effects, and promoting biodiversity.

3. **Social Cohesion:** Community gardens and farmers' markets serve as focal points for community engagement and social interaction, fostering a sense of belonging and collective ownership among residents.

4. **Educational Opportunities:** Urban farming initiatives provide educational opportunities for children and adults alike, teaching valuable skills related to gardening, nutrition, and environmental stewardship.

Urban farming initiatives represent innovative solutions to the complex challenges facing urban areas today. By promoting sustainable food production, enhancing community resilience, and fostering environmental stewardship, these initiatives offer a pathway towards more livable, equitable, and resilient cities. Continued investment and support for urban farming are essential to realizing its full potential in creating thriving urban communities. Exploring Urban Farming Initiatives: Nurturing Sustainable Solutions in Urban Landscapes Urban farming, often synonymous with urban agriculture, stands as a multifaceted strategy to confront the myriad challenges plaguing urban areas today. This discourse aims to delve into urban farming initiatives, underscoring their significance, objectives, and profound impact on urban communities.

Significance of Urban Farming Initiatives

In an era marked by escalating urbanization, urban farming initiatives have emerged as pivotal agents of change. With cities grappling with issues like food insecurity, food deserts, and environmental degradation, urban farming presents a sustainable remedy. By repurposing underutilized urban spaces for food cultivation, these initiatives curtail dependence on long-haul food transportation, thereby mitigating the environmental toll linked with conventional agriculture.

Objectives of Urban Farming Initiatives

Urban farming initiatives are propelled by a spectrum of objectives, including:

1. **Food Security:** Anchored in the ethos of local production, urban farming endeavors to bolster access to fresh, nutritious fare, especially in marginalized urban pockets devoid of adequate supermarket access.

2. **Environmental Sustainability:** Embracing sustainable agricultural practices like organic farming and water conservation, urban farming endeavors to shrink environmental footprints, thereby safeguarding natural resources.

3. **Community Engagement:** Through platforms like community gardens, rooftop farms, and farmers' markets, urban farming initiatives catalyze community engagement, nurturing social cohesion, and providing avenues for education and communal bonding.

4. **Economic Opportunities:** Urban farming burgeons as a source of livelihood, incubating economic prospects for small-scale farmers, entrepreneurs, and ancillary businesses operating within the food sector.

Types of Urban Farming Initiatives

Urban farming initiatives embody diverse modalities, including:

1. **Rooftop Gardens:** Leveraging the rooftops of buildings, rooftop

gardens burgeon as bastions of greenery, fostering food production while augmenting urban aesthetics.

2. **Community Gardens:** Serving as communal sanctuaries, community gardens act as crucibles for collective cultivation endeavors, fostering bonds and promoting wholesome dietary habits.

3. **Vertical Farms:** Harnessing vertical spaces within urban edifices, vertical farms harness hydroponic or aeroponic technologies to engender year-round crop production, regardless of spatial constraints.

4. **Aquaponic Systems:** Marrying aquaculture with hydroponics, aquaponic systems orchestrate a symbiotic relationship between fish and plants, fostering resource efficiency and minimizing water consumption.

Impact of Urban Farming Initiatives

Urban farming initiatives leave an indelible imprint on urban landscapes, engendering:

1. **Enhanced Access to Fresh Produce:** By translocating food production closer to urbanites, urban farming initiatives democratize access to fresh, locally sourced produce, heralding improved dietary habits and bolstering public health.

2. **Environmental Stewardship:** From composting to rainwater harvesting, urban farming initiatives champion eco-friendly practices, curbing greenhouse gas emissions, tempering urban heat island effects, and fostering biodiversity.

3. **Social Cohesion:** Serving as epicenters of communal conviviality, community gardens and farmers' markets foster social interconnectedness, forging bonds and nurturing collective ownership among residents.

4. **Educational Enrichment:** Urban farming initiatives serve as pedagogical conduits, imparting invaluable lessons in gardening, nutrition, and environmental stewardship to denizens of all ages.

Urban farming initiatives epitomize innovative panaceas to the labyrinthine challenges afflicting urban habitats.

By espousing sustainable food production, fortifying community resilience, and championing environmental stewardship, these initiatives chart a trajectory toward more habitable, equitable, and resilient urban milieus. Sustained investment and advocacy for urban farming are imperative to harnessing its full transformative potential in sculpting vibrant urban ecosystems.

Case Studies

Edible Routes (New Delhi, India): Edible Routes is a pioneering urban farming initiative in New Delhi that promotes sustainable food production through various projects, including rooftop gardens, vertical gardens, and community farming. They offer workshops, consultations, and installations to encourage individuals and organizations to grow their own food in urban environments, fostering food security and environmental sustainability.

Green Souls (Mumbai, India): Green Souls is a Mumbai-based organization that specializes in setting up urban farms and green spaces in residential complexes, corporate offices, and educational institutions. Their projects include rooftop gardens, terrace farms, and indoor hydroponic systems, aimed at promoting green living and fostering a connection with nature in the bustling cityscape of Mumbai.

The Terrace Farming Project (Chennai, India): The Terrace Farming Project is an initiative in Chennai that promotes terrace gardening as a means of sustainable food production in urban areas. They provide training, resources, and support to individuals and communities interested in setting up their own terrace gardens, contributing to local food resilience and ecological conservation.

Jaivik Setu (Bangalore, India): Jaivik Setu is a non-profit organization based in Bangalore that facilitates organic farming initiatives in urban and peri-urban areas. They collaborate with local communities, schools, and businesses to establish organic gardens, composting units, and biodiversity zones, promoting healthy living and environmental stewardship in India's Silicon Valley. Sundara Wellness (Pune, India): Sundara Wellness is a wellness center in Pune that incorporates urban farming into its holistic approach to health and wellness. They have established rooftop gardens and green spaces within their facility, where visitors can participate in gardening workshops, yoga sessions, and organic farming activities, fostering a deeper connection with nature and sustainable living practices.

Urban farming initiatives wield a profound influence on community health and well-being, fostering a symbiotic relationship between urban denizens and their local ecosystems. Here's a deeper dive into the manifold ways in which urban farming engenders positive impacts

1. **Access to Fresh, Nutritious Food:** Urban farming endeavors democratize access to fresh, locally sourced produce, thereby mitigating the scourge of food deserts and fortifying community health. By cultivating a diverse array of fruits, vegetables, and herbs within urban locales, these initiatives infuse vitality into urban diets, paving the way for improved nutrition and enhanced well-being among residents.

2. **Promotion of Physical Activity:** Engagement in urban farming activities necessitates physical exertion, be it tilling the soil, planting seeds, or tending to crops. This incidental exercise fosters physical activity, promoting cardiovascular health and mitigating the sedentary lifestyles endemic to urban settings. Moreover, the act of nurturing plants fosters a profound connection with nature, stimulating mental rejuvenation and bolstering overall vitality.

3. **Social Cohesion and Interaction:** Community gardens, rooftop farms, and farmers' markets burgeon as epicenters of communal conviviality, catalyzing social interaction and fostering bonds among urban dwellers. As denizens converge to cultivate, harvest, and share the fruits of their labor, they forge a sense of collective ownership and camaraderie, nurturing a vibrant tapestry of social

cohesion within urban landscapes.

4. **Psychological Well-being:** Immersion in urban farming endeavors confers myriad psychological benefits, ranging from stress reduction to enhanced mental well-being. Studies evince that engagement in gardening activities correlates with diminished levels of cortisol, the stress hormone, thereby fostering emotional equilibrium and psychological resilience amidst the rigors of urban life. Moreover, the therapeutic aspects of horticultural therapy offer solace and sanctuary to individuals grappling with mental health challenges, serving as a balm for the soul amidst urban tumult.

5. **Educational Opportunities:** Urban farming initiatives serve as veritable crucibles of experiential learning, affording individuals of all ages the opportunity to glean insights into agriculture, nutrition, and environmental sustainability. Through hands-on cultivation endeavors, denizens acquire practical skills and ecological literacy, thereby fostering a culture of environmental stewardship and ecological mindfulness within urban communities.

Urban farming stands as a potent catalyst for community health and well-being, fostering a holistic tapestry of physical, social, and psychological flourishing within urban milieus. By nurturing access to fresh, nutritious fare, promoting physical activity and social interaction, and offering fertile grounds for experiential learning, urban farming initiatives emerge as veritable bulwarks against the exigencies of urban living, nurturing resilient, vibrant, and flourishing communities.

Expanding on policy and planning considerations for urban agriculture involves several key aspects:

1. **Integration into City Planning and Zoning Regulations:** Urban agriculture should be integrated into city planning and zoning regulations to ensure that sufficient space is allocated for various forms of agriculture, including community gardens, rooftop farms, vertical gardens, and urban green spaces. This integration requires collaboration between urban planners, policymakers, and community stakeholders to identify suitable locations for agricultural activities within the urban landscape.

2. **Allocation of Space:** City planners should identify and allocate space for urban agriculture within the urban environment. This may involve repurposing vacant lots, unused rooftops, or underutilized spaces for agricultural purposes. Additionally, incorporating green infrastructure into urban development plans can help create more resilient and sustainable cities while providing space for agriculture.

3. **Financial Incentives:** Governments can provide financial incentives to support urban farmers and community gardeners. This may include grants, subsidies, tax incentives, and low-interest loans to help cover startup costs, purchase equipment, or invest in infrastructure improvements. Financial support can also be directed towards research and development initiatives aimed at advancing urban agriculture practices and technologies.

4. **Technical Assistance:** Urban farmers and community gardeners may require technical assistance to navigate regulatory requirements, obtain necessary permits, and access resources. Governments can provide technical support through extension services, workshops, training programs, and networking opportunities. Collaborating with agricultural experts, universities, and research institutions can help ensure that urban farmers have

access to the knowledge and expertise needed to succeed.

5. **Land Access:** Access to land is a significant barrier for many aspiring urban farmers. Governments can address this challenge by identifying and making publicly owned land available for urban agriculture purposes. Additionally, policies such as land trusts, land leasing programs, and zoning regulations that prioritize agricultural use can help secure land for urban farming initiatives.

6. **Infrastructure Development:** Investing in infrastructure such as irrigation systems, composting facilities, and cold storage facilities can help support the growth of urban agriculture. Governments can allocate funding for infrastructure development projects and collaborate with private sector partners to build essential infrastructure that benefits urban farmers and enhances the viability of urban agriculture ventures.

7. **Education and Outreach:** Educating the public about the benefits of urban agriculture and promoting community involvement in food production are essential for the long-term success of urban farming initiatives. Governments can support education and outreach efforts through public awareness campaigns, school programs, and community events that highlight the importance of urban agriculture for food security, environmental sustainability, and community well-being.

By incorporating this policy and planning considerations, policymakers and urban planners can create an enabling environment for the expansion of urban agriculture, thereby promoting food security, environmental sustainability, and economic development in cities.

Agricultural Robotics

Expanding on the introduction to robotic applications in agriculture and examples of robots used for planting involves delving deeper into the functionalities, benefits, and challenges associated with these advanced technologies.

Functionality of Planting Robots:

Planting robots, such as those developed by Naïo Technologies and Small Robot Company, are equipped with advanced technologies that enable them to perform precise and efficient planting tasks. These robots typically utilize a combination of the following functionalities:

1. **Autonomous Navigation:** Planting robots are equipped with GPS and computer vision systems that allow them to navigate fields autonomously. Using real-time data and mapping algorithms, these robots can identify the optimal planting paths and navigate around obstacles in the field.

2. **Precision Planting:** Planting robots employ precision planting techniques to ensure accurate seed placement. They are equipped with planting modules that can adjust seed spacing, depth, and placement according to crop-specific requirements and agronomic

practices. This precision planting capability helps optimize seed distribution and maximize crop yield potential.

3. **Sensor Systems**: Many planting robots are equipped with sensor systems that enable them to gather data about soil conditions, moisture levels, and other environmental factors. These sensors provide valuable insights that can inform planting decisions and optimize seed placement for optimal germination and crop growth.

4. **Integration with Farm Management Systems:** Planting robots are often integrated with farm management systems, allowing farmers to plan, monitor, and analyze planting activities remotely. Integration with these systems enables real-time data exchange, task scheduling, and performance tracking, facilitating better decision-making and operational efficiency.

Benefits of Planting Robots:

The adoption of planting robots offers several benefits to farmers and agricultural operations:

5. **Labor Savings:** Planting robots reduce the reliance on manual labor for planting operations, helping to address labor shortages and reduce labor costs for farmers.

6. **Increased Efficiency:** Planting robots can plant seeds more quickly and accurately than traditional planting methods, leading to

improved efficiency and productivity in farming operations.

7. **Optimized Seed Placement:** Precision planting capabilities ensure that seeds are placed at the optimal spacing and depth, promoting uniform germination and crop establishment for higher yields.

8. **Reduced Environmental Impact:** By minimizing soil disturbance and optimizing seed placement, planting robots help reduce soil erosion, conserve water, and minimize the use of fertilizers and pesticides, contributing to sustainable farming practices.

9. **Enhanced Crop Management:** Planting robots can be integrated with other agricultural technologies, such as crop monitoring systems and predictive analytics, to provide comprehensive crop management solutions. By collecting and analyzing data about crop health, growth, and performance, planting robots help farmers make informed decisions and optimize crop production.

Challenges and Considerations:

Despite their potential benefits, the adoption of planting robots in agriculture also presents several challenges and considerations:

1. **Cost:** Planting robots can represent a significant investment for farmers, including upfront costs for equipment purchase, maintenance, and operation. Assessing the return on investment and long-term cost-effectiveness of these technologies is essential for adoption.

2. **Technology Integration:** Integrating planting robots into existing farming operations and workflows may require adjustments and training for farmers and farm workers. Ensuring seamless integration with other farm management systems and technologies is critical for maximizing the benefits of these robots.

3. **Regulatory Compliance:** Compliance with regulations and safety standards for autonomous vehicles and agricultural machinery is essential for the deployment and operation of planting robots. Ensuring compliance with regulatory requirements and addressing any legal or liability concerns is necessary for widespread adoption.

4. **Data Management and Privacy:** Planting robots generate large amounts of data about planting activities, field conditions, and crop performance. Farmers must implement robust data management practices to protect data privacy, ensure data security, and comply with data regulations.

By addressing these challenges and considerations, farmers and agricultural stakeholders can harness the potential of planting robots to optimize planting operations, improve crop productivity, and advance sustainable agriculture practices.

Expanding on examples of robots used for harvesting and monitoring crops involves providing further details on their functionalities, benefits, and the impact they have on agricultural practices.

Harvesting Robots:

1. **Strawberry-Picking Robot by Octinion:** This robot is specifically designed for harvesting strawberries, a delicate and labor-intensive crop. Equipped with advanced vision systems and robotic arms, the robot can identify ripe strawberries and gently pick them without damaging the fruit or the plant. By automating the harvesting process, the Octinion robot significantly reduces the need for manual labor and increases harvesting efficiency, allowing farmers to harvest their crops in a timely manner and minimize post-harvest losses.

2. **Lettuce-Harvesting Robot by Harvest CROO Robotics:** Designed for harvesting lettuce, this robot utilizes a combination of sensors, cameras, and robotic arms to identify and pick mature heads of lettuce. The robot is capable of navigating between rows of lettuce plants and selectively harvesting only the mature heads, leaving the rest of the crop undisturbed. By automating the labor-intensive task of lettuce harvesting, the Harvest CROO Robotics robot helps reduce labor costs and improve harvesting efficiency, ensuring that farmers can harvest their crops at the peak of freshness and quality.

Monitoring Crop Health Robots:

1. **Blue River See & Spray Robot:** Developed by Blue River Technology (acquired by John Deere), this robot is designed for precision weed control in row crops such as soybeans, cotton, and corn. Equipped with advanced computer vision and machine learning algorithms, the robot can identify weeds in real-time and precisely apply herbicides only where needed, minimizing chemical usage and reducing the environmental impact of weed control. By targeting weeds with surgical precision, the Blue River See & Spray robot helps farmers improve crop yields, reduce input costs, and adopt more sustainable farming practices.

2. **Drones with Multispectral Cameras:** Drones equipped with multispectral cameras are increasingly used for monitoring crop health and detecting early signs of stress, disease, or pest infestations. These drones can capture high-resolution images of crops from above, allowing farmers to analyze vegetation indices, detect anomalies, and make informed decisions about crop management. By providing timely and accurate information about crop health, drones help farmers optimize inputs, minimize losses, and maximize yields, ultimately improving farm profitability and sustainability.

These examples demonstrate how robotics and automation are transforming various aspects of agriculture, from harvesting to crop monitoring, and enabling farmers to improve efficiency, productivity, and sustainability in their operations. As technology continues to advance, the role of robots in agriculture is expected to grow, offering new opportunities for innovation and growth in the agricultural sector.

Expanding on the potential impact of agricultural robotics on labor and productivity involves exploring the implications for farmers, agricultural workers, and the overall agricultural industry.

Labor Reduction and Reallocation:

1. **Reduction of Labor Requirements:** Agricultural robotics can

significantly reduce the need for manual labor in farming operations, particularly for repetitive and physically demanding tasks such as planting, harvesting, and weeding. By automating these tasks, robots can perform them more efficiently and with higher precision than human laborers, thereby reducing labor costs for farmers.

2. **Reallocation of Human Resources:** As robots take on more of the repetitive and labor-intensive tasks in agriculture, human workers can be reallocated to more skilled and strategic roles that require creativity, problem-solving, and decision-making. This shift allows farmers to leverage human expertise in areas such as crop management, pest control, and farm planning, leading to more informed and efficient farming practices.

Increased Productivity and Efficiency:

1. **Continuous Operation:** Agricultural robots can operate 24/7, allowing farming operations to continue around the clock and maximizing productivity, especially during critical periods such as planting and harvesting seasons. Unlike human laborers who have limited working hours, robots can work continuously without the need for breaks or rest, thereby increasing overall farm efficiency.

2. **Precision and Accuracy:** Robots equipped with advanced sensors, computer vision, and machine learning algorithms can perform tasks with greater precision and accuracy than human laborers. For example, planting robots can precisely place seeds at optimal spacing and depth, while harvesting robots can selectively pick ripe fruits and vegetables without damaging the crops. This precision helps optimize crop yield and quality, leading to higher productivity and profitability for farmers.

Enhanced Farm Resilience:

1. **Weather and Environmental Resilience:** Agricultural robots are not affected by adverse weather conditions, such as extreme heat, cold, or rain, which can limit the ability of human laborers to work

in the field. By operating autonomously and without human intervention, robots can continue working under challenging weather conditions, ensuring that farming operations remain resilient and on schedule.

2. **Hazardous Environments:** Some farming tasks, such as applying pesticides or working in confined spaces, pose health and safety risks to human laborers. Agricultural robots can operate in hazardous environments without risking human health, thereby reducing the likelihood of accidents and injuries on the farm. This improves workplace safety and reduces liability for farmers.

Overall, the adoption of agricultural robotics has the potential to transform farming practices, leading to a more efficient, productive, and sustainable agricultural industry. By reducing labor requirements, increasing productivity, and enhancing farm resilience, robots enable farmers to overcome labor challenges, optimize resource utilization, and meet the growing demand for food in a rapidly changing world.

Challenges with robotics

Expanding on the challenges and considerations in adopting agricultural robotics involves exploring the complexities involved in integrating these advanced technologies into farming operations.

High Initial Costs:

1. **Capital Investment:** The initial costs of acquiring agricultural robotics systems can be substantial for farmers, including expenses for purchasing equipment, installation, and setup. The high upfront investment required may pose a barrier to adoption, particularly for small and medium-sized farms with limited financial resources.

2. **Return on Investment (ROI):** Assessing the return on investment for agricultural robotics systems is crucial for farmers to justify the upfront costs. While robotics can offer long-term benefits such as labor savings, increased productivity, and improved crop yields, farmers need to evaluate the potential ROI against the initial investment and ongoing operational costs.

Technical Expertise and Training:

1. **Technical Skills Requirement:** Operating and maintaining agricultural robotics systems often requires specialized technical skills and expertise. Farmers and farm workers may need training to effectively utilize robotics technology, troubleshoot issues, and optimize system performance. Access to training programs and technical support is essential to ensure that farmers can maximize the value of their investment in robotics.

Job Displacement and Social Equity:

1. **Labor Displacement Concerns:** The adoption of agricultural robotics has raised concerns about job displacement and the potential impact on rural communities dependent on agricultural labor. While robots can automate repetitive and labor-intensive tasks, they may also reduce the demand for manual labor in farming operations, leading to workforce displacement and economic challenges for agricultural workers.

2. **Social Equity Considerations:** Ensuring equitable access to agricultural robotics technology is essential to prevent widening disparities between large-scale commercial farms and smallholder or family-owned farms. Policies and initiatives aimed at promoting inclusive access to robotics technology, training, and support can help mitigate social equity concerns and ensure that all farmers can benefit from technological advancements.

Interoperability and Compatibility:

1. **Integration with Existing Systems:** Agricultural robotics systems need to be compatible with existing farm equipment, machinery, and management systems to ensure seamless integration into farming operations. Compatibility issues can arise due to differences in software platforms, data formats, and communication protocols, requiring interoperability standards and solutions to facilitate connectivity and data exchange between different systems.

2. **Data Management and Security:** Agricultural robotics generates large volumes of data about farm operations, crop performance, and environmental conditions. Farmers must implement robust data management and security practices to protect sensitive information, comply with data regulations, and prevent unauthorized access or misuse of data.

Addressing these challenges and considerations requires collaborative efforts from farmers, technology providers, policymakers, and other stakeholders in the agricultural industry. By addressing barriers to adoption and promoting responsible deployment of agricultural robotics, stakeholders can unlock the full potential of these technologies to transform farming practices, improve efficiency, and ensure sustainable food production for the future.

Biotechnology and Genetic Engineering

Expanding on the overview of biotechnology in agriculture involves exploring the diverse applications, benefits, and implications of biotechnological advancements in crop improvement and agricultural sustainability.

Applications of Biotechnology in Agriculture:

1. **Genetic Engineering:** Genetic engineering techniques, such as recombinant DNA technology, allow scientists to introduce specific genes into plant genomes to confer desirable traits. For example, genes can be inserted to enhance pest and disease resistance, improve tolerance to abiotic stresses such as drought and salinity, or increase nutrient content in crops.

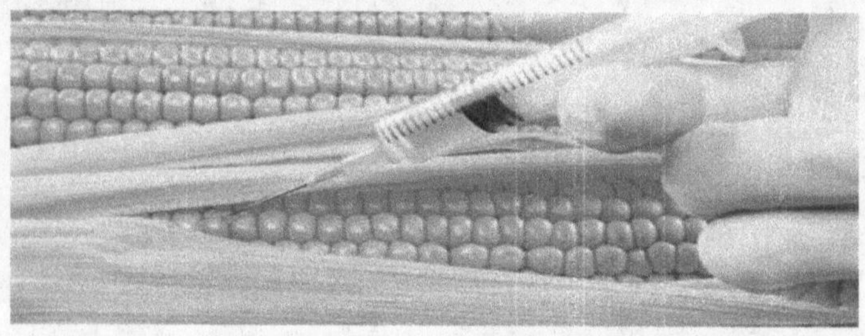

2. **Molecular Breeding:** Molecular breeding techniques enable the identification and selection of desirable traits at the molecular level, facilitating the development of crop varieties with improved yield, quality, and resilience. Marker-assisted selection (MAS) and genomic selection (GS) are examples of molecular breeding approaches used to accelerate the breeding process and enhance genetic gain.

3. **Biological Control:** Biotechnology is utilized in biological control methods to manage pests and diseases in agriculture. Biopesticides derived from microorganisms or natural substances offer

environmentally friendly alternatives to chemical pesticides, reducing the ecological impact of pest management practices.

4. **Plant Tissue Culture:** Plant tissue culture techniques enable the propagation of plants from small tissue samples under controlled conditions. This technology is used for clonal propagation, somatic embryogenesis, and micropropagation of elite plant varieties, allowing for rapid multiplication of plant material with desirable traits.

Benefits of Biotechnology in Agriculture:

1. **Increased Crop Productivity:** Biotechnology has contributed to the development of crop varieties with enhanced productivity, resilience, and resource use efficiency. Traits such as pest resistance, disease tolerance, and drought tolerance help farmers mitigate yield losses and improve crop performance under challenging environmental conditions.

2. **Improved Nutritional Quality:** Biotechnological interventions have led to the development of biofortified crops with improved nutritional content, such as vitamin-enriched rice and iron-fortified beans. These crops address nutritional deficiencies and contribute to improved human health and well-being, particularly in regions where dietary micronutrient deficiencies are prevalent.

3. **Reduced Environmental Impact:** By reducing reliance on

chemical inputs such as pesticides and fertilizers, biotechnology promotes sustainable agricultural practices and minimizes environmental pollution and degradation. Pest-resistant and herbicide-tolerant crop varieties enable farmers to adopt integrated pest management (IPM) and conservation agriculture techniques, resulting in reduced chemical usage and enhanced ecosystem health.

Challenges and Considerations:

1. **Regulatory Framework:** The regulation of biotechnology in agriculture varies across countries and regions, with stringent safety and environmental assessments required for the commercialization of genetically modified (GM) crops. Regulatory frameworks must balance safety considerations with innovation and market access to ensure responsible deployment of biotechnological innovations.

2. **Public Perception and Acceptance:** Public perception of biotechnology in agriculture can influence consumer acceptance and market uptake of biotech-derived products. Addressing concerns related to food safety, environmental impact, and ethical considerations is essential for building public trust and fostering

acceptance of biotechnological advancements.

3. Socioeconomic Implications: Biotechnology may have socioeconomic implications for farmers, consumers, and other stakeholders in the agricultural value chain. Access to biotech-derived crop varieties, intellectual property rights, and distribution of benefits must be managed to ensure equitable outcomes and promote inclusive agricultural development.

Overall, biotechnology plays a crucial role in addressing global agricultural challenges and advancing sustainable food production systems. Continued research, innovation, and collaboration are essential for harnessing the full potential of biotechnology to meet the growing demands for food, feed, and fiber genetically modified (GM) crops and their benefits:

1. **Bt Cotton:** Engineered to produce insecticidal proteins from Bacillus thuringiensis (Bt), Bt cotton reduces the need for chemical insecticides, leading to lower production costs, reduced environmental impact, and increased yields for farmers.

2. **Golden Rice:** Genetically engineered to produce beta-carotene, a precursor of vitamin A, golden rice aims to address vitamin A deficiency, improving nutrition and reducing related illnesses like blindness and immune deficiencies.

3. **Drought-Tolerant Maize:** Genetically modified to withstand water scarcity and drought stress, drought-tolerant maize offers improved yield stability and resilience under adverse environmental conditions, helping farmers maintain productivity in drought-prone regions.

4. **Bt Eggplant (Brinjal):** Engineered to produce Bt proteins toxic to pests like fruit and shoot borers, Bt eggplant reduces the need for chemical insecticides, enhancing pest management and crop protection.

5. **Herbicide-Tolerant Soybean, Corn, and Cotton:** These crops tolerate specific herbicides, simplifying weed management, reducing tillage, conserving soil moisture, and promoting soil health through no-till or reduced-till farming.

6. **Insect-Resistant Corn (Maize):** Producing Bt proteins toxic to insect pests, insect-resistant corn reduces yield losses due to insect damage, increases crop yields, and improves profitability for farmers.

7. **Virus-Resistant Papaya:** Resistant to papaya ringspot virus (PRSV), virus-resistant papaya varieties control PRSV infection, reduce yield losses, and ensure the sustainability of papaya

production in affected regions.

These examples demonstrate how GM crops offer agronomic and socioeconomic benefits, including enhanced pest resistance, improved nutritional quality, increased yield stability, and reduced environmental impact. However, their adoption is subject to regulatory oversight and considerations regarding safety, environmental impact, and socioeconomic implications.

In addition to the environmental and safety concerns surrounding genetic engineering in agriculture, there are several ethical considerations that merit attention:

1. **Environmental Impact:** One of the primary ethical concerns surrounding genetic engineering in agriculture is the potential environmental impact of GM crops. Critics argue that the release of genetically modified organisms (GMOs) into the environment may have unintended consequences, such as the disruption of ecosystems and harm to non-target organisms, including beneficial insects and wildlife. There are also concerns about the long-term effects of GM crops on biodiversity and ecosystem resilience.

2. **Corporate Control:** Critics of genetic engineering express concer about the concentration of corporate control over seed supply and agricultural biotechnology. Large biotechnology companies that develop and market GM crops may hold significant power and influence in shaping agricultural practices and policies, raising questions about equity, farmer autonomy, and access to genetic resources.

3. **Genetic Contamination:** There is a risk of genetic contamination of non-GM crops through cross-pollination with GM crops, particularly in regions where both conventional and GM crops are cultivated. Genetic contamination can compromise the integrity of organic and non-GM supply chains, leading to economic losses for farmers and undermining consumer choice and food sovereignty.

4. **Food Safety:** Ethical concerns related to food safety center on the

potential risks of consuming GM foods. While regulatory agencies typically conduct safety assessments to evaluate the risks associated with GM crops for human health, some consumers may remain skeptical or apprehensive about the safety of GM foods. Ensuring transparency, robust safety evaluations, and accurate labeling of GM products is essential to address consumer concerns and uphold the right to informed choice.

5. **Consumer Choice and Transparency:** There are ethical considerations surrounding consumer choice and transparency in labeling GM products. Some consumers may prefer to avoid GM foods for personal, cultural, or ethical reasons and have the right to make informed choices about the foods they purchase and consume. Transparent labeling that accurately identifies the presence of GM ingredients in food products enables consumers to exercise their right to choose and aligns with principles of autonomy and informed consent.

Addressing these ethical considerations requires a multifaceted approach that involves robust risk assessments, transparent regulatory processes, stakeholder engagement, and ongoing dialogue among policymakers, scientists, farmers, consumers, and civil society organizations. By integrating ethical considerations into decision-making processes and promoting transparency, accountability, and responsible innovation, stakeholders can navigate the complexities of genetic engineering in agriculture while upholding ethical principles and promoting the common good.

Biotechnology has revolutionized agriculture, offering innovative solutions to enhance crop productivity, improve nutritional content, and mitigate environmental challenges. Techniques such as genetic engineering, molecular breeding, and marker-assisted selection are used to modify plant traits and develop crop varieties with desirable characteristics.

Examples of genetically modified (GM) crops include Bt cotton, golden rice, drought-tolerant maize, Bt eggplant, herbicide-tolerant soybean, corn, and cotton, insect-resistant corn, and virus-resistant papaya. These crops offer benefits such as reduced need for chemical inputs, increased yield stability, improved nutritional quality, and reduced environmental impact.

Despite the potential benefits, there are ethical considerations and concerns surrounding genetic engineering in agriculture. These include environmental impact, corporate control over seed supply, genetic contamination, food safety, consumer choice, and transparency in labeling GM products. Regulatory frameworks, stakeholder engagement, and public outreach efforts are essential for addressing these concerns and fostering trust, transparency, and informed decision-making.

Public perception and communication play a significant role in shaping attitudes towards GM crops and influencing regulatory decisions. Effective communication strategies that engage stakeholders and provide accurate information about the safety, benefits, and regulatory oversight of GM crops are crucial for fostering trust and transparency.

Biotechnology and genetic engineering offer promising solutions to address global challenges in agriculture. By balancing the potential benefits and risks of biotechnology and promoting responsible innovation, we can create a more resilient, sustainable, and equitable food system for future generations.

Public perception and communication are crucial aspects that influence attitudes toward genetically modified (GM) crops and ultimately impact regulatory decisions. Effective communication strategies are essential for engaging stakeholders across various sectors, including farmers, consumers, scientists, policymakers, and civil society organizations. These strategies can foster trust, transparency, and informed decision-making about biotechnology in agriculture.

Public outreach and education initiatives are valuable tools for dispelling misconceptions and providing accurate information about GM crops. These efforts can help address concerns related to safety, benefits, and regulatory oversight. By offering accessible and scientifically sound information, public outreach programs contribute to a better understanding of the complexities surrounding GM crops and their role in agriculture.

Stakeholder engagement processes are another critical component of effective communication in the context of GM crops. By involving diverse perspectives and values, stakeholders can contribute to constructive dialogue and consensus-building on complex issues. Engaging with stakeholders ensures that their concerns and perspectives are heard and considered in decision-making processes related to biotechnology and genetic engineering in agriculture.

Overall, fostering open and transparent communication is essential for building public trust and confidence in GM crops. By promoting dialogue, providing accurate information, and engaging stakeholders, we can navigate the complexities of biotechnology in agriculture and work towards sustainable and inclusive solutions for food security and agricultural sustainability.

Importance of Soil Health in Sustainable Agriculture

Cover Cropping:

1. **Soil Erosion Control:** Cover crops help prevent soil erosion by shielding the soil surface from the impact of raindrops and wind. Their dense root systems hold soil particles in place, reducing the risk of erosion and maintaining soil structure.

2. **Nutrient Cycling**: Cover crops play a vital role in nutrient cycling by capturing nutrients from deeper soil layers and making them available to subsequent crops. When cover crops decompose, they release nutrients like nitrogen, phosphorus, and potassium back into the soil, enhancing soil fertility and reducing the need for external inputs.

3. **Weed Management:** Certain cover crops have allelopathic properties, releasing biochemical compounds that inhibit weed germination and growth. By suppressing weed growth, cover crops

help reduce the reliance on herbicides and manual weed control methods, contributing to sustainable weed management practices.

4. **Improved Soil Biology:** Cover crops support a diverse microbial community in the soil, including beneficial bacteria and fungi that contribute to nutrient cycling, disease suppression, and soil aggregation. By providing a habitat and food source for soil organisms, cover crops enhance soil biological activity and resilience.

No-Till Farming:

1. **Soil Moisture Conservation:** No-till farming improves soil structure and aggregation, increasing water infiltration and reducing surface runoff. This leads to better soil moisture retention, especially during dry periods, which can enhance crop resilience to drought stress.

2. **Energy Conservation:** By eliminating or reducing tillage operations, no-till farming conserves energy and reduces fuel consumption associated with tractor use. This can lead to cost savings for farmers and lower greenhouse gas emissions from agricultural machinery.

3. **Biodiversity Promotion:** No-till farming creates a more stable and diverse habitat for soil organisms, including earthworms, beneficial microbes, and arthropods. These organisms play critical roles in soil health and ecosystem functioning, contributing to overall biodiversity in agricultural landscapes.

4. **Long-Term Soil Health:** Over time, no-till farming builds soil organic matter levels and improves soil structure, resulting in healthier and more productive soils. This can lead to increased crop yields, reduced input costs, and greater resilience to environmental stresses.

Overall, both cover cropping and no-till farming are valuable tools for sustainable soil management, offering multiple benefits for soil health, crop productivity, and environmental sustainability. When combined with other conservation practices such as crop rotation and integrated pest management, these practices can contribute to the development of resilient and regenerative agricultural systems that support both people and the planet.

Case studies showcasing successful implementations of regenerative agriculture:

1. Gabe Brown's Ranch, North Dakota, USA:

Gabe Brown, a farmer and rancher in North Dakota, USA, has become a leading figure in the regenerative agriculture movement. Through his innovative approach to farming, Brown has transformed his ranch into a model of soil health and biodiversity. Key practices implemented on Brown's ranch include:

Cover Cropping: Brown incorporates diverse cover crops into his rotations to improve soil health, suppress weeds, and enhance nutrient cycling.

Crop Rotation: By diversifying his crop rotation, Brown reduces pest and disease pressure, improves soil structure, and increases overall resilience.

Holistic Grazing Management: Brown employs rotational grazing strategies to mimic natural grazing patterns, promoting soil fertility, biodiversity, and carbon sequestration.

The results of Brown's regenerative practices have been remarkable. He has seen improvements in soil fertility, water infiltration rates, and biodiversity on his farm. Furthermore, Brown has achieved higher crop yields, reduced input costs, and improved resilience to extreme weather events, demonstrating the economic viability of regenerative agriculture.

2. Singh Farms, Punjab, India:

Singh Farms, located in Punjab, India, is another success story in regenerative agriculture. Facing challenges of soil degradation and water scarcity, the Singh family has adopted regenerative practices to restore soil health and improve water management. Key strategies implemented at Singh Farms include:

Diversified Cropping Systems: The farm has shifted from monoculture to diversified cropping systems, including intercropping and crop rotation, to enhance soil fertility and resilience.

Agroforestry: Singh Farms incorporates trees and shrubs into their agricultural landscapes to improve soil structure, enhance biodiversity, and provide additional sources of income.

Integrated Livestock: Livestock are integrated into cropping systems, providing natural fertilizer, controlling weeds, and contributing to nutrient cycling.

Singh Farms has witnessed significant improvements in soil health, water retention, and crop productivity since adopting regenerative agriculture practices. Additionally, the farm has become a demonstration site for neighboring farmers, showcasing the benefits of regenerative agriculture and inspiring others to make similar transitions.

These case studies illustrate the transformative potential of regenerative agriculture in restoring soil health, enhancing ecosystem resilience, and promoting sustainable land management practices. By prioritizing soil health and adopting regenerative agriculture principles, farmers can build resilient and productive agricultural systems that support food security, environmental sustainability, and rural livelihoods.

Climate Change

Climate-smart agriculture (CSA) encompasses various practices and strategies designed to enhance agricultural productivity, resilience to climate change, and sustainability while reducing greenhouse gas emissions. By addressing the interconnected challenges of agriculture, climate change, and sustainable development, CSA aims to ensure food security and environmental sustainability in the face of climate variability and extreme weather events.

Strategies for Mitigating the Impacts of Climate Change on Farming:

1. **Crop Diversification:** Diversifying crops helps mitigate risks associated with climate variability. By planting a variety of crops with different growth patterns and tolerance levels, farmers can reduce vulnerability to droughts, floods, and pest outbreaks. Crop diversification spreads risk and maintains productivity in changing climate conditions.

2. **Water Management:** Improving water management practices is crucial for mitigating the impacts of water scarcity and droughts on farming. Techniques such as rainwater harvesting, drip irrigation, and soil moisture conservation enhance water use efficiency. Adopting climate-resilient crops and planting strategies further optimizes water usage, reducing reliance on irrigation.

3. **Soil Conservation:** Implementing soil conservation practices is essential for maintaining soil health and resilience to climate change. Practices such as conservation tillage, cover cropping, and agroforestry enhance soil fertility, reduce erosion, and sequester carbon in the soil. Healthy soils provide a stable foundation for sustainable agriculture and are more resilient to climate variability and extreme weather events.

Examples of Climate-Smart Agriculture Initiatives Worldwide:

1. **Conservation Agriculture in Zambia:** The Conservation Farming Unit (CFU) in Zambia promotes conservation agriculture practices such as minimum tillage, crop rotation, and mulching. These practices improve soil fertility, conserve water, and enhance crop resilience to climate change. As a result, farmers adopting these practices have increased yields, reduced production costs, and improved food security.

2. **Climate-Resilient Rice Farming in Vietnam:** In Vietnam's Mekong Delta, farmers are adopting climate-resilient rice farming techniques to address challenges such as saltwater intrusion and flooding. These techniques include early maturing rice varieties, raised bed planting, and alternate wetting and drying irrigation. By adapting to climate change, farmers maintain rice productivity while reducing greenhouse gas emissions and conserving water resources.

3. **Agroforestry Systems in Kenya:** Agroforestry systems, such as the Farmer Managed Natural Regeneration (FMNR) approach in Kenya, integrate trees and crops on farmland to enhance soil fertility, biodiversity, and resilience to climate change. By restoring degraded landscapes and diversifying income sources, agroforestry systems improve farm productivity and livelihoods while promoting environmental sustainability.

These examples illustrate the diverse approaches to climate-smart agriculture worldwide, highlighting the importance of context-specific solutions tailored to local conditions and challenges. By adopting climate-smart practices, farmers can build resilience, enhance sustainability, and contribute to global efforts to address climate change while ensuring food security and livelihoods for future generations.

Blockchain Technology

Blockchain technology is making waves in agriculture, offering solutions to enhance transparency, security, and traceability throughout the food supply chain. Here's an expanded overview:

Blockchain in Agriculture:

Originally developed for cryptocurrency transactions, blockchain technology has evolved into a versatile tool with applications in various industries, including agriculture. In the agricultural sector, blockchain serves as a decentralized and immutable ledger that records transactions and data in a secure and transparent manner.

Key Features of Blockchain:

1. **Decentralization:** Blockchain operates on a decentralized network of computers, known as nodes, which collectively verify and record transactions. This eliminates the need for a central authority, reducing the risk of data manipulation or fraud.

2. **Immutability:** Once data is recorded on the blockchain, it cannot be altered or tampered with, ensuring the integrity and reliability of the information stored within the ledger.

3. **Transparency:** Blockchain provides transparent access to transaction records for all participants in the network. This transparency fosters trust among stakeholders and enables greater visibility into the movement of agricultural products along the supply chain.

4. **Traceability:** Each transaction recorded on the blockchain is timestamped and linked to previous transactions, creating a chronological chain of data. This enables precise tracking of agricultural products from farm to fork, facilitating rapid identification of the source of contamination or quality issues.

Applications of Blockchain in Agriculture:

1. **Supply Chain Management:** Blockchain technology enables end-

to-end visibility and traceability of agricultural products throughout the supply chain. By recording each stage of production, processing, and distribution on the blockchain, stakeholders can track the journey of food products, ensuring compliance with quality standards and regulatory requirements.

2. **Quality Assurance:** Blockchain can be used to authenticate the origin and quality of agricultural products, providing consumers with assurance regarding product authenticity, freshness, and safety. For example, blockchain-enabled QR codes on product packaging allow consumers to access detailed information about the product's journey from farm to table.

3. **Smart Contracts:** Smart contracts, self-executing contracts with the terms of the agreement directly written into code, can automate various processes in agriculture, such as payments, deliveries, and quality control. By leveraging blockchain technology, smart contracts ensure transparency, efficiency, and trust in contractual agreements between farmers, suppliers, and buyers.

4. **Data Management:** Blockchain facilitates secure and transparent management of agricultural data, including crop yield, soil health, and weather conditions. By securely storing and sharing data on the blockchain, farmers, researchers, and policymakers can collaborate more effectively and make data-driven decisions to improve agricultural productivity and sustainability.

Challenges and Considerations:

While blockchain offers significant potential benefits for agriculture, there are challenges and considerations that must be addressed for widespread adoption. These include scalability issues, interoperability with existing systems, data privacy concerns, and the need for standardized protocols and regulations. Additionally, educating stakeholders about the benefits and limitations of blockchain technology is crucial for fostering trust and adoption within the agricultural community. Overall, blockchain technology holds promise for revolutionizing the agricultural sector by enhancing transparency, security, and traceability throughout the food supply chain. By leveraging blockchain-based solutions, stakeholders can build more resilient, efficient, and sustainable agricultural systems that meet the demands of a rapidly changing world.

Case Studies of Blockchain Implementation in the Agricultural Supply Chain**

1. **Walmart's Blockchain Pilot:** In 2018, Walmart partnered with IBM to pilot a blockchain-based system for tracking and tracing fresh produce in its supply chain. The system, known as the IBM Food Trust platform, allows Walmart to trace the journey of produce, such as mangoes and pork, from farm to store in real-time. By leveraging blockchain technology, Walmart aims to enhance food safety, improve supply chain efficiency, and reduce food waste.

2. **Tea Traceability in Malawi:** The Malawi Tea 2020 project, led by the Ethical Tea Partnership (ETP) and agricultural technology firm Bext360, utilizes blockchain technology to trace the origin and quality of tea in Malawi. By tagging tea sacks with QR codes and collecting data on production and quality parameters, stakeholders can track the journey of tea from smallholder farms to processing facilities. This traceability system promotes transparency, sustainability, and fair trade practices in the Malawian tea industry.

3. **India's Coffee Supply Chain:** The Coffee Board of India, in collaboration with Eka Software Solutions, has implemented a blockchain-based system to enhance transparency and traceability in the Indian coffee supply chain. Through this initiative, coffee beans are tagged with QR codes at various stages of production, from cultivation to processing and export. The blockchain platform records information such as origin, quality, and certifications, allowing stakeholders to track the journey of coffee beans and verify their authenticity. This transparent and traceable supply chain promotes fair trade practices, improves market access for farmers,

and enhances consumer trust in Indian coffee products.

3. **Maersk's TradeLens Platform for Agri-Exports:** Maersk, a global logistics company, has developed the TradeLens platform in collaboration with IBM to digitize and streamline global trade processes, including agricultural exports. By leveraging blockchain technology, TradeLens provides a secure and transparent platform for tracking shipments of agricultural products from farm to port. This digital solution enhances visibility and efficiency in the supply chain, reducing paperwork, delays, and errors associated with traditional shipping processes. By improving logistics and documentation processes, TradeLens facilitates smoother trade transactions and enables exporters to deliver fresh agricultural products to international markets more efficiently.

4. **Provenance in Indonesian Palm Oil Industry:** Provenance, a UK-based technology company, has partnered with palm oil producers in Indonesia to enhance transparency and sustainability in the palm oil supply chain. Through the use of blockchain technology, Provenance enables palm oil producers to track the journey of palm oil from plantation to processing facilities and ultimately to consumers. By providing a transparent record of production practices, certifications, and social and environmental impacts, Provenance empowers consumers to make informed purchasing decisions and encourages sustainable practices in the palm oil industry. This traceability solution fosters trust and accountability among stakeholders, driving positive change towards more ethical and sustainable palm oil production.

These case studies illustrate the diverse applications of blockchain technology in the agricultural supply chain, from enhancing traceability and transparency to improving efficiency and sustainability. As more organizations adopt blockchain solutions, we can expect to see further innovation and collaboration to address key challenges in the agricultural sector.

The Farm-to-Table Movement: Cultivating Sustainability and Community Connections

The farm-to-table movement has emerged as a powerful force in reshaping the way we think about food, agriculture, and community. Rooted in principles of sustainability, environmental stewardship, and local resilience, this movement champions the idea of bringing consumers closer to the source of their food, fostering connections between farmers, producers, and consumers, and promoting a deeper understanding of the food system.

At its core, the farm-to-table movement prioritizes the sourcing of food from local farmers, ranchers, and food producers. By supporting local agriculture, consumers not only reduce their carbon footprint by minimizing the distance food travels from farm to plate but also contribute to the vitality of local economies and communities. This emphasis on locality promotes biodiversity, preserves farmland, and strengthens the social fabric by fostering relationships between farmers and consumers.

Seasonal eating is another key tenet of the farm-to-table movement, encouraging consumers to embrace the natural rhythms of the seasons and consume foods that are fresh, in-season, and locally available. By choosing seasonal produce, consumers not only enjoy superior flavor and nutritional quality but also support sustainable farming practices and reduce the need for long-distance transportation and refrigeration.

Transparency and traceability are fundamental values of the farm-to-table movement, ensuring that consumers have access to information about the origins, production methods, and handling practices of the food they consume. This transparency fosters trust and accountability in the food system, empowering consumers to make informed choices about the foods they eat and the impact of their dietary decisions on their health, the environment, and the broader community.

The farm-to-table movement is not just about food—it's about building connections, fostering community, and celebrating the richness and diversity of local food cultures. Farmers' markets, community-supported agriculture (CSA) programs, and farm-to-table restaurants serve as vibrant hubs of activity, bringing together farmers, producers, chefs, and consumers around shared values of sustainability, health, and social responsibility.

As we look to the future, the farm-to-table movement holds immense potential to transform our food system, making it more resilient, equitable, and sustainable for generations to come. By continuing to support local agriculture, embrace seasonal eating, and advocate for transparency and traceability in the food system, we can cultivate a healthier, more connected world—one plate at a time.

Challenges Facing the Farm-to-Table Movement

1. **Access and Affordability:** While the farm-to-table movement promotes the consumption of fresh, locally sourced foods, access to such products can be limited for individuals living in food deserts or low-income communities. The higher cost of locally produced goods, compared to mass-produced alternatives, can present a

barrier to entry for many consumers. Addressing issues of affordability and expanding access to fresh, healthy foods is essential for ensuring that the benefits of the farm-to-table movement are accessible to all.

2. **Infrastructure and Distribution:** Developing robust infrastructure for local food distribution is critical for the success of the farm-to-table movement. Many small-scale farmers lack access to distribution networks and cold storage facilities needed to bring their products to market efficiently. Investing in infrastructure improvements, such as shared-use commercial kitchens, mobile markets, and refrigerated storage facilities, can help bridge the gap between producers and consumers, enabling more farmers to participate in local food systems.

3. **Education and Awareness:** Promoting consumer education and awareness is key to overcoming misconceptions and barriers to adopting farm-to-table practices. Many individuals are unfamiliar with the benefits of seasonal eating, the importance of supporting local agriculture, and the environmental impacts of conventional food systems. By providing educational resources, hosting cooking demonstrations, and engaging with the community through outreach initiatives, advocates of the farm-to-table movement can empower consumers to make informed choices about their food purchases.

Innovations in the Farm-to-Table Movement

1. **Technology Integration:** The integration of technology into the farm-to-table movement has the potential to revolutionize how food is produced, distributed, and consumed. Online platforms, mobile apps, and blockchain technology can enhance transparency and traceability in the food supply chain, allowing consumers to access detailed information about the origin and journey of their food products. Additionally, precision agriculture technologies, such as drones and sensors, can improve farm efficiency and

productivity, enabling farmers to produce more food with fewer resources.

2. **Urban Agriculture:** Urban agriculture initiatives are emerging as a viable solution to increasing food access and promoting local food production in urban areas. Rooftop gardens, community gardens, and vertical farming systems allow city dwellers to grow fresh produce in their own neighborhoods, reducing the distance food travels from farm to table. Urban agriculture not only provides access to healthy foods but also strengthens community bonds and revitalizes urban spaces.

3. **Food Hubs and Aggregators:** Food hubs and aggregators play a crucial role in connecting small-scale farmers with wholesale buyers, such as restaurants, schools, and institutions. By aggregating products from multiple farmers and providing distribution services, food hubs help streamline the local food supply chain, making it more efficient and cost-effective for both producers and buyers. Additionally, food hubs often offer value-added services, such as food processing and packaging, further enhancing the marketability of local products.

The Future of the Farm-to-Table Movement

1. **Policy Support:** Policy support at the local, state, and federal levels is essential for advancing the goals of the farm-to-table movement. Governments can incentivize sustainable agriculture practices, provide funding for local food infrastructure projects, and implement policies that promote food sovereignty and food justice. By creating an enabling environment for local food systems to thrive, policymakers can help realize the potential of the farm-to-table movement to create healthier, more resilient communities.

2. **Collaborative Partnerships:** Collaboration between diverse stakeholders, including farmers, chefs, policymakers, educators, and community organizers, is essential for driving systemic change within the food system. By working together to address shared

challenges and leverage collective expertise, stakeholders can create synergies that amplify the impact of individual efforts. Collaborative partnerships can take many forms, from multi-stakeholder initiatives and coalition building to cross-sectoral projects and knowledge sharing networks.

3. **Cultural Transformation:** Achieving lasting change within the food system requires a cultural transformation in how society values and interacts with food. The farm-to-table movement has the potential to shift societal norms around food consumption, encouraging people to prioritize quality over quantity, sustainability over convenience, and community over individualism. By fostering a deeper connection to the land, the people who cultivate it, and the food that sustains us, the farm-to-table movement can inspire a more conscious and compassionate approach to eating and living.

Importance of Local Food Systems and Community-Supported Agriculture (CSA)**

Local food systems and community-supported agriculture (CSA) play a crucial role in promoting food security, environmental sustainability, and community resilience.

1. **Food Security:** Local food systems reduce dependency on global supply chains and ensure access to fresh, nutritious food for communities, especially in rural and underserved areas. By supporting local farmers, consumers contribute to a more resilient and diversified food system that is less vulnerable to disruptions and price fluctuations.

2. **Environmental Sustainability:** Local food systems minimize the carbon footprint associated with food production and distribution by reducing transportation distances and reliance on fossil fuels. Additionally, supporting local agriculture promotes sustainable farming practices that protect soil health, conserve water resources, and preserve biodiversity.

3. **Community Resilience:** Community-supported agriculture

(CSA) programs foster direct relationships between farmers and consumers, creating a sense of connection, trust, and shared responsibility. By participating in CSAs, consumers invest in the success of local farms, support rural economies, and strengthen community ties. Certainly!

4. **Economic Vitality:** Local food systems and CSA programs contribute to the economic vitality of rural and urban communities alike. By keeping food dollars circulating within the local economy, these initiatives create jobs, stimulate entrepreneurship, and support small-scale farmers and food producers. Furthermore, by fostering direct relationships between producers and consumers, local food systems empower farmers to receive fair prices for their products, enabling them to reinvest in their businesses and communities.

5. **Cultural Preservation:** Local food systems celebrate and preserve culinary traditions, regional flavors, and agricultural heritage. By prioritizing locally grown and culturally significant foods, communities can maintain connections to their cultural roots and promote a sense of pride and identity. Additionally, local food systems provide opportunities for cultural exchange and storytelling, as farmers and producers share their knowledge, traditions, and recipes with consumers.

6. **Education and Food Literacy:** Local food systems and CSA programs serve as invaluable educational resources, providing opportunities for individuals to learn about where their food comes from, how it is produced, and the importance of sustainable agriculture. Through farm tours, cooking classes, and hands-on experiences, consumers gain a deeper understanding of the seasonal rhythms of agriculture, the challenges faced by farmers, and the environmental impacts of food choices. This increased food literacy empowers individuals to make informed decisions about their food purchases and cultivates a greater appreciation for the interconnectedness of food, health, and the environment.

7. **Food Justice and Equity:** Local food systems and CSA programs have the potential to advance food justice and equity by addressing issues of food access, affordability, and sovereignty. By prioritizing community needs and engaging marginalized populations, these initiatives can ensure that all individuals have access to fresh, healthy, culturally appropriate foods. Additionally, by supporting diverse and inclusive food systems, local food initiatives can challenge systemic inequalities within the food system and promote social justice and equity.

Local food systems and community-supported agriculture (CSA) are vital components of a sustainable and resilient food system. By promoting food security, environmental sustainability, economic vitality, cultural preservation, education, and food justice, these initiatives contribute to the well-being of individuals, communities, and the planet. Through collective action and collaboration, we can continue to strengthen and expand local food systems, creating a more equitable, resilient, and sustai Examples of Farm-to-Table Initiatives Promoting

Sustainable Food Consumption

1. **Farmers' Markets:** Farmers' markets bring together local farmers, artisans, and food producers to sell their products directly to consumers. These vibrant community hubs offer fresh, seasonal produce, artisanal goods, and prepared foods, creating opportunities for consumers to connect with local farmers and learn about sustainable food production.

2. **Community Gardens:** Community gardens provide space for people to grow their own food, share resources, and build community connections. These grassroots initiatives promote self-sufficiency, food sovereignty, and environmental stewardship, empowering individuals to cultivate healthy, sustainable lifestyles.

3. **Farm-to-School Programs:** Farm-to-school programs connect schools with local farms to serve fresh, locally sourced meals to students and promote nutrition education and environmental awareness. These programs support local agriculture, improve children's access to healthy food, and instill lifelong habits of healthy eating and sustainability.

4. **Community-Supported Agriculture (CSA):** CSA programs allow consumers to purchase shares or subscriptions directly from local farms, receiving a weekly or monthly supply of fresh produce and other farm products. By participating in CSAs, consumers support local farmers financially, share in the risks and rewards of agricultural production, and gain a deeper appreciation for seasonal eating and sustainable agriculture practices.

5. **Farm-to-Table Restaurants:** Farm-to-table restaurants prioritize sourcing ingredients from local farms and food producers, often featuring seasonal menus that highlight the flavors of the

region. These restaurants support local agriculture, celebrate culinary diversity, and promote sustainable food practices by forging direct relationships with farmers and incorporating fresh, locally sourced ingredients into their menus.

5. **Food Cooperatives:** Food cooperatives, or co-ops, are community-owned grocery stores that prioritize sourcing products from local farmers and producers. By pooling resources and purchasing collectively, co-op members support local agriculture, promote food transparency and traceability, and build resilient, self-reliant food systems that prioritize community well-being over profit.

6. **Urban Farming Initiatives:** Urban farming initiatives, such as rooftop gardens, community gardens, and urban agriculture projects, bring food production closer to urban centers, reducing the environmental impact of food transportation and increasing access to fresh, healthy produce in densely populated areas. These initiatives promote food sovereignty, urban greening, and community resilience, empowering residents to take control of their food supply and build more sustainable, equitable food systems.

4. **8.Food Rescue and Redistribution Programs:** Food rescue and redistribution programs collect surplus food from farms,

restaurants, and grocery stores that would otherwise go to waste and redistribute it to those in need. By preventing food waste and addressing food insecurity, these initiatives promote sustainable consumption and support local communities by ensuring that nutritious food is accessible to all, regardless of income or background.

Speculation on Future Innovations in Agriculture

As technology continues to advance and global challenges such as climate change and food insecurity persists, agriculture is poised to undergo significant transformations in the coming decades. Some potential future innovations in agriculture include:

1. **Precision Agriculture 2.0:** Building upon current precision agriculture technologies, future innovations may include the widespread adoption of advanced sensors, drones, and artificial intelligence to optimize resource use, improve decision-making, and increase productivity.

2. **Vertical Farming Expansion:** Vertical farming may become more prevalent, especially in urban areas, as advancements in lighting, automation, and hydroponic systems make indoor agriculture more efficient and economically viable.

3. **Gene Editing and Synthetic Biology:** Continued advancements in gene editing technologies, such as CRISPR-Cas9, may enable the development of crops with enhanced traits such as disease resistance, nutrient efficiency, and environmental resilience.

4. **Smart Farming Systems:** The integration of Internet of Things (IoT) devices and data analytics will enable the development of smart farming systems that monitor environmental conditions, crop health, and machinery performance in real-time. These systems will facilitate proactive decision-making, allowing farmers to optimize inputs, minimize waste, and maximize yields.

5. **Climate-Resilient Crops:** With climate change posing increasing challenges to agriculture, there will be a growing emphasis on developing climate-resilient crop varieties that can thrive in changing environmental conditions, such as drought, heat,

and salinity. Breeding programs and genetic engineering techniques will play a crucial role in creating crops that are better adapted to future climatic scenarios.

6. **Regenerative Agriculture Practices:** Regenerative agriculture practices, such as agroforestry, cover cropping, and rotational grazing, will gain traction as farmers seek sustainable ways to improve soil health, sequester carbon, and enhance biodiversity on their lands. These holistic farming approaches will not only mitigate climate change but also increase resilience to extreme weather events and enhance ecosystem services.

7. **Blockchain in Supply Chain Management:** Blockchain technology will revolutionize supply chain management in agriculture by providing transparent and immutable records of transactions, from farm to fork. By enhancing traceability, authenticity, and food safety, blockchain will enable consumers to make more informed choices about the products they buy and incentivize ethical and sustainable practices throughout the supply chain.

8. **Urban Agriculture Integration:** Urban agriculture will become increasingly integrated into urban planning and development strategies, with rooftop gardens, vertical farms, and community gardens becoming common features of city landscapes. These urban agriculture initiatives will enhance food security, promote local food production, and create green spaces that improve air quality and mitigate urban heat island effects.

Challenges Facing Agriculture

Despite the promise of future innovations, agriculture faces several significant challenges that must be addressed to ensure food security, environmental sustainability, and rural livelihoods. 1. **Population Growth:** With the global population projected to exceed 9 billion by 2050, agriculture will need to produce more food with fewer resources while minimizing environmental impact.

Climate Change**

Climate change presents one of the most pressing challenges to agriculture, with its impacts already being felt globally. Extreme weather events, such as droughts, floods, heatwaves, and storms, disrupt agricultural operations, reduce yields, and threaten food security. Additionally, shifting temperature and precipitation patterns alter growing seasons and crop suitability, requiring farmers to adapt their practices and crop choices. Rising temperatures also exacerbate pest and disease pressures, further jeopardizing crop yields. Mitigating climate change through emission reductions and sequestering carbon in agricultural soils is crucial for building resilience and ensuring the long-term viability of agriculture.

Resource Depletion

Resource depletion, including soil degradation, water scarcity, and loss of biodiversity, poses significant threats to agricultural sustainability. Soil erosion, nutrient depletion, and salinization degrade soil quality, reduce fertility, and diminish productivity, undermining the ability of farmers to meet growing food demands. Water scarcity, exacerbated by climate change and overexploitation of aquifers, restricts irrigation and threatens crop yields, particularly in arid and semi-arid regions. Loss of biodiversity, including pollinators and natural predators, disrupts ecosystem services essential for pest control and soil fertility. Implementing sustainable land management practices, such as conservation agriculture, agroforestry, and integrated pest management, is essential for restoring and preserving natural resources and ecosystems.

Food Insecurity and Rural Poverty

Despite advances in agricultural productivity, food insecurity remains a persistent challenge, with millions of people worldwide lacking access to sufficient, nutritious food. Poverty, inequality, conflict, and inadequate infrastructure exacerbate food insecurity, particularly in rural areas where the majority of the world's poor and hungry reside. Smallholder farmers, who often lack access to resources, technology, and markets, are disproportionately affected by food insecurity and rural poverty. Strengthening rural livelihoods through investments in agricultural infrastructure, market access, education, and social protection programs is essential for reducing poverty, enhancing food security, and fostering sustainable development.

Recommendations for Policymakers**

1. ****Investment in Research and Innovation:**** Policymakers should prioritize funding for agricultural research and development to support the development and adoption of sustainable farming practices, climate-resilient crops, and technological innovations.

2. ****Policy Support for Sustainable Agriculture:**** Policies that incentivize sustainable farming practices, conservation efforts, and climate-smart agriculture can help address environmental challenges while promoting economic growth and rural development.

3. ****Promotion of Access to Resources:**** Policymakers should prioritize policies that ensure equitable access to land, water, seeds, and other agricultural resources, particularly for smallholder farmers and marginalized communities. This includes measures to protect land rights, improve access to credit and inputs, and support agroecological approaches that enhance resource efficiency and

resilience.

4. **Investment in Rural Infrastructure:** Infrastructure investment, including roads, storage facilities, irrigation systems, and market access, is critical for improving agricultural productivity, reducing post-harvest losses, and connecting farmers to markets. Policymakers should prioritize investments in rural infrastructure to support agricultural growth, market integration, and rural development.

5. **Support for Farmer Education and Extension Services:** Access to knowledge, training, and extension services is essential for empowering farmers to adopt sustainable practices, improve productivity, and adapt to changing environmental conditions. Policymakers should invest in farmer education programs, extension services, and farmer-led research initiatives to build capacity, disseminate best practices, and promote innovation at the grassroots level.

6. **Strengthening Market Institutions:** Policymakers should work to strengthen market institutions, improve market transparency, and promote fair trade practices to ensure that farmers receive fair prices for their products and have access to markets. This includes measures to promote local food systems, support farmers' cooperatives and collective marketing initiatives, and enforce regulations to prevent market distortions and exploitation.

7. **Climate Change Adaptation and Mitigation:** Policymakers should prioritize climate change adaptation and mitigation efforts within the agricultural sector, including measures to enhance resilience to climate impacts, reduce greenhouse gas emissions, and promote carbon sequestration in agricultural soils. This may involve supporting agroecological practices, promoting agroforestry and soil conservation measures, and incentivizing climate-smart agricultural investments.

8. **Integration of Food Security and Nutrition Goals:**

Policymakers should integrate food security and nutrition goals into agricultural policies and programs, recognizing the interconnectedness of food systems, health, and livelihoods. This includes measures to improve access to diverse and nutritious foods, promote dietary diversity, and address underlying drivers of malnutrition, such as poverty and inequality.

By implementing these recommendations, policymakers can help build a more sustainable, equitable, and resilient agricultural sector that addresses the challenges of food security, environmental sustainability, and rural development while promoting inclusive growth and prosperity for all.

Recommendations for Farmers

1. **Adoption of Sustainable Practices:** Farmers should prioritize sustainable farming practices such as conservation agriculture, agroforestry, and integrated pest management to improve soil health, conserve resources, and reduce environmental impact.

2. **Diversification and Resilience:** Diversifying crops and income sources can help farmers mitigate risks associated with climate variability, market fluctuations, and pest outbreaks, improving resilience and long-term viability.

3. **Investment in Soil Health:** Farmers should prioritize investments in soil health through practices such as cover cropping, crop rotation, and minimal tillage. Healthy soils are essential for maintaining fertility, water retention, and nutrient cycling, supporting crop productivity and resilience to climate extremes.

4. **Water Management:** Efficient water management is critical for sustainable agriculture, particularly in regions facing water scarcity or drought. Farmers should adopt water-saving technologies such as drip irrigation, rainwater harvesting, and soil moisture monitoring to optimize water use efficiency and minimize irrigation water losses.

5. **Climate-Smart Agriculture:** Farmers should embrace climate-smart agriculture techniques that help adapt to and mitigate climate

change impacts. This includes selecting climate-resilient crop varieties, adjusting planting dates, and implementing water-saving measures to cope with changing weather patterns and extreme events.

6. **Integrated Pest and Disease Management:** Farmers should employ integrated pest and disease management strategies that minimize reliance on chemical pesticides and promote natural pest control methods. This includes crop rotation, biological pest control, and the use of resistant crop varieties to reduce pest pressure while preserving ecosystem health.

7. **Continuous Learning and Adaptation:** Agriculture is constantly evolving, and farmers should prioritize continuous learning, experimentation, and adaptation to stay abreast of emerging trends, technologies, and best practices. Participating in farmer-to-farmer networks, attending training workshops, and leveraging digital agriculture tools can help farmers access valuable knowledge and resources to improve their operations.

8. **Market Diversification and Value-Added Products:** Farmers should explore opportunities to diversify their market channels and add value to their products through processing, packaging, and branding. Direct marketing avenues such as farmers' markets, community-supported agriculture (CSA) programs, and farm-to-table restaurants can provide premium prices and direct connections with consumers, enhancing farm profitability and sustainability.

9. **Farm Financial Management:** Sound financial management is essential for farm viability and resilience. Farmers should develop comprehensive farm budgets, manage cash flow effectively, and explore risk management tools such as crop insurance and hedging to mitigate financial risks associated with farming operations.

10. **Collaboration and Networking:** Collaboration with other farmers, agricultural experts, and stakeholders is invaluable for sharing knowledge, accessing resources, and advocating for policy support. Participating in farmer cooperatives, producer associations,

and agribusiness networks can provide opportunities for collective action, joint marketing, and mutual support within the farming community.

By embracing these recommendations, farmers can enhance the sustainability, resilience, and profitability of their operations while contributing to broader goals of environmental stewardship, food security, and rural development.

Recommendations for Consumers**

1. ****Support Local and Sustainable Food Systems:**** Consumers can support local farmers and sustainable food systems by purchasing locally sourced, seasonal foods, participating in community-supported agriculture (CSA) programs, and advocating for policies that promote food sovereignty and environmental stewardship.

2. ****Reduce Food Waste:**** Consumers can minimize food waste by practicing mindful consumption, meal planning, and composting organic waste, reducing pressure on landfills and conserving resources.

3. ****Choose Nutrient-Dense Foods:**** Consumers can prioritize nutrient-dense foods such as fruits, vegetables, whole grains, and lean proteins to support their health and well-being while minimizing the environmental impact of their diets. Choosing minimally processed foods and incorporating plant-based options into meals can also reduce the carbon footprint associated with food production and transportation.

4. ****Support Ethical and Transparent Brands:**** Consumers can support ethical and transparent food brands by seeking out products that prioritize fair labor practices, animal welfare, and environmental sustainability. Certifications such as Fair Trade, USDA Organic, and Certified Humane provide assurance that products meet certain standards of ethical and sustainable production.

5. ****Reduce Meat and Dairy Consumption:**** Consumers can reduce their consumption of meat and dairy products, which have significant

environmental impacts due to their resource-intensive production processes and greenhouse gas emissions. Choosing plant-based alternatives, participating in Meatless Mondays, and incorporating more plant-based meals into their diets can help reduce the environmental footprint of food consumption.

6. **Educate Yourself:** Consumers can educate themselves about where their food comes from, how it is produced, and the impacts of their food choices on health, the environment, and society. Reading labels, researching brands, and staying informed about food-related issues can empower consumers to make more informed and ethical purchasing decisions.

7. **Advocate for Change:** Consumers can advocate for policies and initiatives that promote food justice, environmental sustainability, and public health. This includes supporting legislation that addresses food insecurity, promotes regenerative agriculture practices, and protects vulnerable communities from the impacts of climate change and industrial agriculture.

8. **Engage in Food Community:** Consumers can engage with their local food community by participating in farmers' markets, community gardens, and food cooperatives, fostering connections with local farmers, producers, and fellow consumers. By supporting local food initiatives and community-building efforts, consumers can contribute to a more resilient, equitable, and sustainable food system.

The journey through the pages of this book has unveiled a plethora of remarkable advances in agriculture that have transformed the landscape of food production, sustainability, and efficiency. From precision farming techniques to genetic engineering breakthroughs, each chapter has offered insights into the innovative strategies driving this vital industry forward.

The pivotal role of technology in revolutionizing agricultural practices, precision agriculture, enabled by advancements in sensors, drones, and data analytics, has empowered farmers to optimize resource allocation, minimize waste, and enhance crop yields. Similarly, genetic engineering tools like CRISPR-Cas9 have opened new frontiers in crop improvement, offering solutions to challenges such as pest resistance, drought tolerance, and nutritional enhancement.

Moreover, the emphasis on sustainability and environmental stewardship has been a recurring motif throughout this exploration. Sustainable agriculture practices, including organic farming, agroforestry, and conservation tillage, have gained prominence as society grapples with the urgent need to mitigate climate change and preserve biodiversity. The integration of renewable energy sources, such as solar and wind power, into agricultural operations further underscores the commitment to reducing carbon emissions and fostering resilience in the face of environmental pressures.

As we reflect on these achievements, it becomes evident that the future of agriculture holds immense promise and potential. However, challenges such as climate variability, resource scarcity, and socio-economic disparities loom large on the horizon. Addressing these complex issues will require interdisciplinary collaboration, policy innovation, and a concerted effort to ensure that the benefits of agricultural advancement are equitably distributed across diverse communities.

Looking ahead, it is imperative that we continue to foster a culture of innovation and knowledge sharing within the agricultural community. By embracing emerging technologies, promoting sustainable practices, and prioritizing inclusivity and diversity, we can build a resilient and thriving agricultural system that nourishes both people and the planet.

There is need of ingenuity, perseverance, and collective effort of scientists, farmers, policymakers, and stakeholders who are committed to shaping a more sustainable and prosperous future for agriculture. As we turn the final page, let us carry forward the lessons learned and continue this transformative journey with renewed determination and optimism.